ADAPTIVE FUZZY SYSTEMS AND CONTROL

Design and Stability Analysis

Li-Xin Wang

University of California at Berkeley

PTR Prentice Hall
Englewood Cliffs, New Jersey 07632

Library of Congress Cataloging in Publication Data

Wang, Li-Xin, (date)
    Adaptive fuzzy systems and control: design and stability analysis
    / Li-Xin Wang.
    p.   cm.

    Includes bibliographical references and index.
    ISBN 0-13-099631-9
    1. Adaptive Control Systems. 2. Fuzzy systems. I. Title.
TJ217.W15 1994   93-25570
629.8'36--dc20   CIP

Acquisitions editor: *Karen Gettman*
Editorial/production supervision: *Kim Gueterman*
Cover design: *Doug DeLuca*
Production coordinator: *Alexis Heydt*

© 1994 by PTR Prentice Hall
Prentice-Hall, Inc.
A Paramount Communications Company
Englewood Cliffs, New Jersey 07632

All rights reserved. No part of this book may be
reproduced, in any form or by any means,
without permission in writing from the publisher.

Printed in the United States of America
10 9 8 7 6 5 4 3 2 1

ISBN 0-13-099631-9

Prentice-Hall International (UK) Limited, *London*
Prentice-Hall of Australia Pty. Limited, *Sydney*
Prentice-Hall Canada Inc., *Toronto*
Prentice-Hall Hispanoamericana, S.A., *Mexico*
Prentice-Hall of India Private Limited, *New Delhi*
Prentice-Hall of Japan, Inc., *Tokyo*
Simon & Schuster Asia Pte. Ltd., *Singapore*
Editora Prentice-Hall do Brasil, Ltda., *Rio de Janeiro*

To Yingbi

ML

UNIVERSITY OF STRATHCLYDE

30125 00471215 3

Books are to be returned on or before
the last date below.

26 JUL 2011

AY 2000

ADAPTIVE FUZZY SYSTEMS AND CONTROL

CONTENTS

FOREWORD xi

PREFACE xii

1 INTRODUCTION 1

    1.1   Combining Numerical and Linguistic Information into Engineering Systems—An Adaptive Fuzzy System Approach 1

    1.2   Classification of Fuzzy Logic Systems 2

          1.2.1 Pure Fuzzy Logic Systems 3
          1.2.2 Takagi and Sugeno's Fuzzy System 3
          1.2.3 Fuzzy Logic Systems with Fuzzifier and Defuzzifier 5

    1.3   Outline of This Book 6

PART I Analysis and Design of Adaptive Fuzzy Systems

2 DESCRIPTION AND ANALYSIS OF FUZZY LOGIC SYSTEMS 9

    2.1   Basic Concepts of Fuzzy Sets and Fuzzy Logic 9

        2.1.1 Fuzzy Set and Set-Theoretical Operators 9
        2.1.2 The Extension Principle 11
        2.1.3 Fuzzy Relations and Their Compositions 11
        2.1.4 Linguistic Variables and Hedges 12
        2.1.5 Generalized Modus Ponens and Generalized Modus Tollens 13
        2.1.6 Fuzzy Implications 14

2.2   Fuzzy Rule Base 15

2.3   Fuzzy Inference Engine 17

        2.3.1 Interpretations of a Fuzzy IF-THEN Rule 17
        2.3.2 Properties of Interpretations 19
        2.3.3 Overall Mapping of the Fuzzy Inference Engine 21

2.4   Fuzzifier 22

2.5   Defuzzifier 22

2.6   Useful Classes of Fuzzy Logic Systems 23

        2.6.1 How to Select Appropriate Operators in the Fuzzy Logic Systems 23
        2.6.2 Derivation of Useful Classes of Fuzzy Logic Systems 24

2.7   Fuzzy Logic Systems as Universal Approximators 27

2.8   Concluding Remarks 28

3   TRAINING OF FUZZY LOGIC SYSTEMS USING BACK-PROPAGATION 29

3.1   Introduction 29

3.2   Back-Propagation Training Algorithm for the Fuzzy Logic Systems 29

3.3   Application to Nonlinear Dynamic System Identification 32

        3.3.1 Motivation 32
        3.3.2 Conceptual Advantages of the Fuzzy Identifiers Over the Neural Identifiers 32
        3.3.3 Design of the Fuzzy Identifiers 33
        3.3.4 Simulations 37

3.4   Concluding Remarks 47

Contents   ix

4   TRAINING OF FUZZY LOGIC SYSTEMS USING ORTHOGONAL LEAST SQUARES   49

4.1   Introduction   49

4.2   Fuzzy Systems as Fuzzy Basis Function Expansions   50

4.3   Orthogonal Least Squares Learning   52

4.4   Control of the Nonlinear Ball-and-Beam System Using FBF Expansions   54

4.5   Modeling the Mackey-Glass Chaotic Time Series by FBF Expansion   60

4.6   Concluding Remarks   64

5   TRAINING OF FUZZY LOGIC SYSTEMS USING A TABLE-LOOKUP SCHEME   65

5.1   Introduction   65

5.2   Generating Fuzzy Rules from Numerical Data   65

5.3   Application to Truck Backer-Upper Control   69

5.4   Application to Time Series Prediction   76

5.5   Concluding Remarks   82

6   TRAINING OF FUZZY LOGIC SYSTEMS USING NEAREST NEIGHBORHOOD CLUSTERING   83

6.1   Introduction   83

6.2   An Optimal Fuzzy Logic System   84

6.3   An Adaptive Version of the Optimal Fuzzy Logic System   85

6.4   Application to Adaptive Control of Nonlinear Dynamic Systems   87

6.5   Concluding Remarks   91

7 COMPARISON OF ADAPTIVE FUZZY SYSTEMS WITH ARTIFICIAL NEURAL NETWORKS 93

7.1 Introduction 93

7.2 Comparison of Multilayer Perceptron with Adaptive Fuzzy Systems 94

7.3 Comparison of Radial Basis Function Networks with Adaptive Fuzzy Systems 96

7.4 Comparison of Probabilistic General Regression with Adaptive Fuzzy Systems 98

7.5 Concluding Remarks 100

PART II Adaptive Fuzzy Control and Filtering

8 STABLE INDIRECT ADAPTIVE FUZZY CONTROL OF NONLINEAR SYSTEMS 102

8.1 Introduction 102

  8.1.1 Why Fuzzy Control? 102
  8.1.2 Why Adaptive Fuzzy Control? 104
  8.1.3 Direct and Indirect Adaptive Fuzzy Control 104
  8.1.4 First and Second Types of Adaptive Fuzzy Control 105

8.2 A Constructive Lyapunov Synthesis Approach to Indirect Adaptive Fuzzy Controller Design 107

  8.2.1 Control Objectives 107
  8.2.2 Certainty Equivalent Controller 108
  8.2.3 Supervisory Control 109
  8.2.4 Adaptive Law 110

8.3 Design and Stability Analysis of First-Type Indirect Adaptive Fuzzy Controllers 114

8.4 Design and Stability Analysis of Second-Type Indirect Adaptive Fuzzy Controllers 118

8.5 Application to Inverted Pendulum Tracking Control 122

8.6 Concluding Remarks 139

9 STABLE DIRECT ADAPTIVE FUZZY CONTROL OF NONLINEAR SYSTEMS 140

9.1 Introduction 140

9.2 Basic Ideas of Constructing Direct Adaptive Fuzzy Controllers 140

9.3 Design and Stability Analysis of First-Type Direct Adaptive Fuzzy Controllers 144

9.4 Design and Stability Analysis of Second-Type Direct Adaptive Fuzzy Controllers 146

9.5 Simulations 148

9.6 Concluding Remarks 153

10 DESIGN OF ADAPTIVE FUZZY CONTROLLERS USING INPUT-OUTPUT LINEARIZATION CONCEPT 155

10.1 Introduction 155

10.2 Intuitive Concepts of Input-Output Linearization 156

10.3 Design of Adaptive Fuzzy Controllers Based on the Input-Output Linearization Concept 158

10.4 Application to the Ball- and-Beam System Control 160

10.5 Concluding Remarks 163

11 DESIGN AND STABILITY ANALYSIS OF FUZZY IDENTIFIERS OF NONLINEAR DYNAMIC SYSTEMS 164

11.1 Introduction 164

11.2 Fuzzy Identifiers as Universal Approximators of Dynamic Systems 165

11.3 Design and Stability Analysis of First-Type Fuzzy Identifiers 166

11.4 Design and Stability Analysis of Second-Type Fuzzy Identifiers 171

11.5 Application to the Chaotic Glycolytic Oscillator Identification 175

11.6 Concluding Remarks 181

12 FUZZY ADAPTIVE FILTERS 182

12.1 Introduction 182

12.2 RLS Fuzzy Adaptive Filter 183

12.3 LMS Fuzzy Adaptive Filter 186

12.4 Application to Nonlinear Channel Equalization 189

12.5 Concluding Remarks 201

13 CONCLUSIONS 202

13.1 General Conclusions 202

13.2 Future Research 204

Appendices

A PROOFS OF THE UNIVERSAL APPROXIMATION AND STABILITY THEOREMS 210

B REFERENCES 221

INDEX 229

FOREWORD

The past three years have witnessed a rapid growth in the use of fuzzy logic in a wide variety of consumer products and industrial systems. Prominent examples of such use are electronically stabilized camcorders, autofocus cameras, washing machines, air conditioners, automobile transmissions, subway trains, and cement kilns.

Despite the visible successes of fuzzy logic, there is still a substantial misunderstanding of what fuzzy logic is, how it compares with other system design methodologies, and what are its strengths and limitations. In part, the misunderstanding stems from a duality of the meaning of fuzzy logic. In its narrow sense, fuzzy logic is a logic of approximate reasoning which may be viewed as a generalization and extension of multivalued logic. But in a broader and much more significant sense, fuzzy logic is coextensive with the theory of fuzzy sets, that is, classes of objects in which the transition from membership to nonmembership is gradual rather than abrupt. In its wider sense—which is becoming predominant in the literature—fuzzy logic has many branches ranging from fuzzy arithmetic and fuzzy automata to fuzzy pattern recognition, fuzzy languages and fuzzy expert systems. In fact, any field X can be fuzzified and called fuzzy X by replacing the concept of a set in X by the concept of a fuzzy set.

What is important to recognize is that fuzzy logic in its narrow sense plays a relatively small role in fuzzy control. Indeed what is essential to the understanding of fuzzy control is the theory of fuzzy relations and, in particular, the calculus of fuzzy if-then rules or, more simply, the calculus of fuzzy rules (CFR). Typically, a fuzzy rule may be expressed as *if X is A then Y is B*, where X and Y are variables and A and B are their linguistic values, for example, *small* and *large*, which are interpreted as labels of fuzzy sets in their respective universes of discourse.

Basically, the calculus of fuzzy rules provides an effective methodology for dealing with imprecise dependencies in systems analysis. The methodology of CFR differs very substantially—both in spirit and in substance— from the conventional approaches which are employed in control theory and systems design. An important feature of CFR which is stressed in Dr. Wang's work is that it is close to human intuition. In fact, most of our experiences are stored in our memory in the form of fuzzy rules and almost all of human reasoning involves a manipulation of such rules on both conscious and subconscious levels. In this perspective, the role model for the calculus of fuzzy rules is the human mind.

Dr. Wang's work lies at the frontiers of the calculus of fuzzy rules and, especially, its application to system identification and adaptive system design. A problem of central importance in CFR is that of the induction of fuzzy rules from observations. Seminal contributions to the formulation and solution of this problem were made by Procyk and Mamdani in 1977 and by Takagi, Sugeno and Kang in 1985 and 1988. The approach developed by Dr. Wang is based on an architecture which is somewhat similar to that employed by Takagi, Sugeno and Kang (TSK). But unlike the approach used by TSK, Dr. Wang makes a skillful use of the techniques developed in neural network theory and, in particular, the backpropagation algorithm. Related to Dr. Wang's approach are important contributions by Kosko, Jang, Lee and Lin, H. Takagi, and others.

What stands out in Dr. Wang's work is the thoroughness of his analysis and his success in formulating a general method which can be applied not only to the conception and design of stable, adaptive fuzzy systems but also to a wide variety of other problems including the solution of systems of algebraic equations, pattern classification and signal processing. The importance of Dr. Wang's contribution is certain to grow with the passage of time and accumulation of experience. However, it should be noted that Dr. Wang's approach as well as similar approaches by other contributors address the problem of the induction of fuzzy rules from input/output data only in the context of multilayer feedforward architectures.

Dr. Wang is a very clear expositor both orally and in writing. As a research monograph, the present volume is an important contribution both to the theory of CFR and its applications. It belongs on the desk of everyone who has a serious interest in systems analysis and intelligent control.

Lotfi A. Zadeh

University of California at Berkeley

PREFACE

A good engineering approach should be capable of making use of all the available information effectively. For many practical problems, an important portion of information comes from human experts. Usually, the expert information is not precise and is represented by fuzzy terms like *small*, *large*, *not very large*, and so on. There are many reasons why expert information is usually expressed in fuzzy terms, such as for convenience or lack of more precise knowledge, ease of communication, and so forth. In order to make use of the expert information in a systematic manner, the so-called intelligent approaches (intelligent control, intelligent signal processing, and others) have been emerging in the engineering community. However, most such intelligent approaches are ad hoc in nature in the sense that there are no analytic tools for general design procedures to guarantee basic performance criteria. Usually, these intelligent approaches combine expert systems with conventional engineering systems in an ad hoc manner, and then simulations are performed to show the validity of the approaches to the specific problems. There are serious limitations to using expert information in this way because it is inefficient, is not generally applicable, and has no guarantee of performance.

An important research topic is to develop general approaches to incorporate expert information systematically for which theoretical analyses can be performed to study the performance of the resulting systems, for example, the stability of an intelligent control system. However, this is not an easy task. In addition to the expert information, another important portion of information is numerical information, which is collected from various sensors or obtained according to physical laws.

Numerical information and expert linguistic information have many fundamental differences. For example, numerical information obeys physical laws and mathematical axioms, whereas there are no such laws and axioms for linguistic information. In other words, the laws governing linguistic information are fundamentally different from the laws governing numerical information. There are two worlds—the physical world and the human world—and as man-machine interaction increases, more and more engineering systems belong to the mixture of these two different worlds. In order to analyze the systems in this mixed world, it is essential that we find a common framework to represent key elements in these two worlds. Adaptive fuzzy systems provide such a common framework.

An adaptive fuzzy system is a fuzzy logic system equipped with a training algorithm, where the fuzzy logic system is constructed from a collection of fuzzy IF-THEN rules, and the training algorithm adjusts the parameters of the fuzzy logic system based on numerical input-output pairs. Conceptually, adaptive fuzzy systems combine linguistic and numerical information in the following way. Because fuzzy logic systems are constructed from fuzzy IF-THEN rules, linguistic information (in the form of fuzzy IF-THEN rules) can be directly incorporated; on the other hand, numerical information (in the form of input-output pairs) is incorporated by training the fuzzy logic system to match the input-output pairs. In this book, we will develop a variety of adaptive fuzzy systems and apply them to a selected number of engineering problems. Throughout this book, we will emphasize how to combine linguistic and numerical information using various kinds of adaptive fuzzy systems and show examples of how the performance is improved by using these adaptive fuzzy system approaches.

Most work in this research monograph was done when I was a Ph.D. student at the University of Southern California. Therefore, I would like to express my first acknowledgment to my Ph.D. advisor, Professor Jerry M. Mendel. Without his constant support and encouragement throughout my Ph.D. study, this book would not exist. Most of all, I would like to thank Professor Mendel for giving me the chance to study in the United States.

I would like to express my special appreciation to my present postdoctoral advisor, Professor Lotfi A. Zadeh for his constant and invaluable encouragement. The final version of this book was written while I was at the University of California at Berkeley. I greatly appreciate the many helpful comments and suggestions resulting from Professor Zadeh's review of the manuscript.

My extended appreciation goes to Professor Shankar Sastry, Professor Charles Desoer, Professor Wei Ren, and Dr. Hideyuki Takagi for their comments on various problems in this book. I would also like to thank the "fuzzy group" here at UC Berkeley—Dr. Jyh-Shing R. Jang, Dr. Chuen-Tsai Sun, Mr. Raymond L. Chen, and Dr. Pratap Khedkar—for helpful discussions on various topics of fuzzy theory. Especially, I would like to thank Dr. Chuen-Chien Lee, a former member of the "fuzzy group," for his excellent tutorial paper [Lee, 33; 34] which introduced me to this fuzzy field.

The support for the author from the Rockwell International Science Center and Berkeley Initiative in Soft Computing is greatly appreciated.

Finally, I would like to express my deepest appreciation to my wife Yingbi. Her constant support and encouragement make my research easier and more enjoyable. She has given so much in order to support me, and she deserves the best gift I could ever give to her—this book.

Li-Xin Wang

University of California at Berkeley

ADAPTIVE FUZZY SYSTEMS AND CONTROL

1

INTRODUCTION

1.1 COMBINING NUMERICAL AND LINGUISTIC INFORMATION INTO ENGINEERING SYSTEMS— AN ADAPTIVE FUZZY SYSTEM APPROACH

For most engineering systems, there are two important information sources: *sensors* which provide numerical measurements of variables, and *human experts* who provide linguistic instructions and descriptions about the system. We call the information from sensors *numerical information* and the information from human experts *linguistic information*. Numerical information is represented by numbers, for example—0.25, 1.44, and so on, whereas linguistic information is represented by words like *small*, *large*, *very large*, and so forth. Conventional engineering approaches can only make use of numerical information and have difficulty incorporating linguistic information. Because so much human knowledge is represented in linguistic terms, incorporating it into engineering systems in a systematic and efficient manner is very important.

Why is linguistic information usually represented in fuzzy terms? We think that there are at least three reasons. First, we usually find it more convenient and efficient to communicate our knowledge in fuzzy terms. This is understandable because if we insist on using only crisp terms, then we must first have precise definitions of these crisp terms. This in turn may result in a chain of definitions—a very inefficient and inconvenient procedure which clearly does not happen in our everyday lives. Second, our knowledge about many problems is essentially fuzzy. For example, when we first learn a new theory, we often find that we understand *something* about the theory, for example, its motivation, basic ideas, advantages, disadvantages, and so on, but we are not sure about some of the details. Now if we are asked to introduce the theory to another person, then that person can only get a fuzzy picture of the theory. The interesting point is that although the picture is not clear, it may serve the purpose quite well—for example, knowing the motivation, basic ideas, advantages, and disadvantages may be sufficient for a higher-level manager. Third, many systems are too complex to describe in crisp terms. For example, our knowledge about a complex chemical process may only be represented in fuzzy terms, for example, "if the temperature is high, then the reaction is intense." The important point here is that although this kind of linguistic information is not precise, it provides important information about the system, and sometimes it may be the only information source. We should make use of this fuzzy information in a scientific way.

The goal of this book is to develop a collection of methods which can effectively combine numerical and linguistic information into engineering systems, to apply them to a variety of control, signal processing, and communication problems, and to analyze their performance (for example, stability). We will use adaptive fuzzy systems as a tool for achieving this goal.

An *adaptive fuzzy system* is defined as a fuzzy logic system equipped with a training algorithm, where the fuzzy logic system is constructed from a set of fuzzy IF-THEN rules using fuzzy logic principles, and the training algorithm adjusts the parameters (and the structures) of the fuzzy logic system based on numerical information. Adaptive fuzzy systems can be viewed as fuzzy logic systems whose rules are automatically generated through training. There are two strategies of combining numerical and linguistic information using adaptive fuzzy systems:

- Use linguistic information to construct an initial fuzzy logic system, and then adjust the parameters of the initial fuzzy logic system based on numerical information. The final fuzzy logic system is, therefore, constructed based on both numerical and linguistic information.
- Use numerical information and linguistic information to construct two separate fuzzy logic systems, and then average them to obtain the final fuzzy logic system.

In order to understand how the preceding procedures work, the first question might be: What is a fuzzy logic system?

1.2 CLASSIFICATION OF FUZZY LOGIC SYSTEMS

Fuzzy logic systems is a name for the systems which have a direct relationship with fuzzy concepts (like fuzzy sets, linguistic variables, and so on) and fuzzy logic.[1] The most popular fuzzy logic systems in the literature may be classified into three

---

[1] Readers unfamiliar with the basic concepts of fuzzy sets and fuzzy logic may read Section 2.1 before reading this section.

types: pure fuzzy logic systems, Takagi and Sugeno's fuzzy system, and fuzzy logic systems with fuzzifier and defuzzifier, which are briefly described in the next three subsections.

1.2.1 Pure Fuzzy Logic Systems

The basic configuration of a pure fuzzy logic system is shown in Figure 1.1 where the fuzzy rule base consists of a collection of fuzzy IF-THEN rules, and the fuzzy inference engine uses these fuzzy IF-THEN rules to determine a mapping from fuzzy sets in the input universe of discourse $U \subset R^n$ to fuzzy sets in the output universe of discourse $V \subset R$ based on fuzzy logic principles. The fuzzy IF-THEN rules are of the following form:

$$R^{(l)} : \text{IF } x_1 \text{ is } F_1^l \text{ and } \cdots \text{ and } x_n \text{ is } F_n^l, \text{ THEN } y \text{ is } G^l \quad (1.1)$$

where F_i^l and G^l are fuzzy sets, $\underline{x} = (x_1, \ldots, x_n)^T \in U$ and $y \in V$ are input and output linguistic variables, respectively, and $l = 1, 2, \ldots, M$. Practice has shown that these fuzzy IF-THEN rules provide a convenient framework to incorporate human experts' knowledge. Each fuzzy IF-THEN rule of (1.1) defines fuzzy set $F_1^l \times \cdots \times F_n^l \to G^l$ in the product space $U \times V$. The most commonly used fuzzy logic principle in the fuzzy inference engine is the so-called sup-star composition. Specifically, let A' be an arbitrary fuzzy set in U; that is, A' is the input to the pure fuzzy logic system of Figure 1.1, then the output determined by each fuzzy IF-THEN rule of (1.1) is a fuzzy set $A' \circ R^{(l)}$ in V whose membership function is

$$\mu_{A' \circ R^{(l)}}(y) = sup_{\underline{x} \in U}[\mu_{A'}(\underline{x}) \star \mu_{F_1^l \times \cdots \times F_n^l \to G^l}(\underline{x}, y)] \quad (1.2)$$

where the "$\star$" operator is "min," "product," or others shown in Chapter 2. We use μ_A to represent the membership function of fuzzy set A. The final output of the pure fuzzy logic system is a fuzzy set $A' \circ (R^{(1)}, \ldots, R^{(M)})$ in V which is a

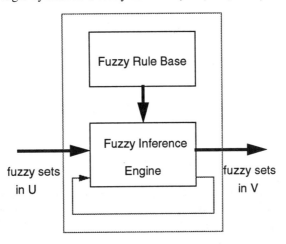

Figure 1.1 Basic configuration of pure fuzzy logic system.

combination of the M fuzzy sets of (1.2); that is,

$$\mu_{A' \circ (R^{(1)},\ldots,R^{(M)})}(y) = \mu_{A' \circ R^{(1)}}(y) \dot{+} \cdots \dot{+} \mu_{A' \circ R^{(M)}}(y) \tag{1.3}$$

where the "$\dot{+}$" operator is "max," "algebraic sum" ($x \dot{+} y = x + y - xy$), or others shown in Chapter 2. If there is feedback as shown by the dashed arrow line in Figure 1.1, we have the so-called fuzzy dynamic systems, that is, pure fuzzy logic systems whose inputs depend on their outputs.

The pure fuzzy logic system constitutes the essential part of fuzzy logic systems. It is a general framework in which linguistic information from human experts is quantified and fuzzy logic principles are used to make systematic use of linguistic information. A main disadvantage of a pure fuzzy logic system is that its inputs and outputs are fuzzy sets, whereas in most engineering systems the inputs and outputs of a system are real-valued variables. To overcome this disadvantage, Takagi and Sugeno [72] proposed another fuzzy logic system whose inputs and outputs are real-valued variables.

1.2.2 Takagi and Sugeno's Fuzzy System

Instead of considering the fuzzy IF-THEN rules in the form of (1.1), Takagi and Sugeno [72] proposed to use the following fuzzy IF-THEN rules:

$$L^{(l)}: \quad \text{IF} \quad x_1 \text{ is } F_1^l \text{ and } \cdots \text{ and } x_n \text{ is } F_n^l,$$
$$\text{THEN} \quad y^l = c_0^l + c_1^l x_1 + \cdots + c_n^l x_n \tag{1.4}$$

where F_i^l are fuzzy sets, c_i are real-valued parameters, y^l is the system output due to rule $L^{(l)}$, and $l = 1, 2, \ldots, M$. That is, they considered rules whose IF part is fuzzy but whose THEN part is crisp—the output is a linear combination of input variables. For a real-valued input vector $\underline{x} = (x_1, \ldots, x_n)^T$, the output $y(\underline{x})$ of Takagi and Sugeno's fuzzy system is a weighted average of the y^l's:

$$y(\underline{x}) = \frac{\sum_{l=1}^{M} w^l y^l}{\sum_{l=1}^{M} w^l} \tag{1.5}$$

where the weight w^l implies the overall truth value of the premise of rule $L^{(l)}$ for the input and is calculated as

$$w^l = \prod_{i=1}^{n} \mu_{F_i^l}(x_i) \tag{1.6}$$

The configuration of Takagi and Sugeno's fuzzy system is shown in Figure 1.2. Takagi and Sugeno's fuzzy system has been successfully applied to many practical problems. The advantage of this fuzzy logic system is that it provides a compact system equation (1.5) and, therefore, parameter estimation and order determination methods can be developed to estimate the parameters c_i^l and the order M. A weak point of this fuzzy logic system is that the THEN part of the rule is not fuzzy; therefore, it does not provide a natural framework to incorporate fuzzy rules from human experts. That is, modification of the pure fuzzy rule (1.1) (for

Sec. 1.2 Classification of Fuzzy Logic Systems

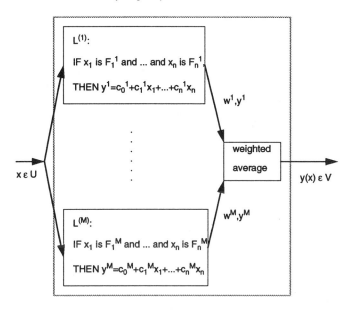

Figure 1.2 Basic configuration of Takagi and Sugeno's fuzzy system.

example, change "y is G^l" to "$y = c_0^l$" where c_0^l is the center of μ_{G^l}) must be performed in order to incorporate it into the fuzzy logic system. Also, there is not much freedom left to apply different principles in fuzzy logic. In this book, we use fuzzy logic systems of the following type.

1.2.3 Fuzzy Logic Systems with Fuzzifier and Defuzzifier

In order to use the pure fuzzy logic system shown in Figure 1.1 in engineering systems where inputs and outputs are real-valued variables, the most straightforward way is to add a fuzzifier to the input and a defuzzifier to the output of the pure fuzzy logic system. The basic configuration of fuzzy logic systems with fuzzifier and defuzzifier is shown in Figure 1.3. The fuzzifier maps crisp points in U to fuzzy sets in U, and the defuzzifier maps fuzzy sets in V to crisp points in V. The fuzzy rule base and fuzzy inference engine are the same as those in the pure fuzzy logic system. In the literature, this fuzzy logic system is often called the fuzzy logic controller since it has been mainly used as a controller. It was first proposed by Mamdani [41] and has been successfully applied to a variety of industrial processes and consumer products. We provide a detailed description of this fuzzy logic system in Chapter 2.

The fuzzy logic system with fuzzifier and defuzzifier has many attractive features. First, it is suitable for engineering systems because its inputs and outputs are real-valued variables. Second, it provides a natural framework to incorporate fuzzy IF-THEN rules from human experts. Third, there is much freedom in the

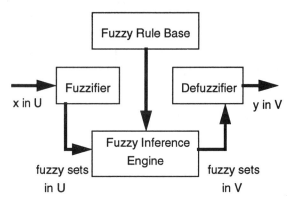

Figure 1.3 Basic configuration of fuzzy logic system with fuzzifier and defuzzifier.

choices of fuzzifier, fuzzy inference engine, and defuzzifier (see Chapter 2 for details), so that we may obtain the most suitable fuzzy logic system for a particular problem. Finally, we can develop different training algorithms for this fuzzy logic system (see Chapters 3–6) so that it provides an effective framework to integrate numerical and linguistic information. In the rest of this book, we only consider the fuzzy logic systems with fuzzifier and defuzzifier.

1.3 OUTLINE OF THIS BOOK

This book consists of two parts. In Part I (Chapters 2–7), we present a detailed description of fuzzy logic systems, develop four training algorithms for the fuzzy logic systems, apply the resulting adaptive fuzzy systems to control and signal processing problems, and compare the adaptive fuzzy systems with artificial neural networks. In Part II (Chapters 8–13), we concentrate on control and filtering applications of adaptive fuzzy systems. We design several adaptive fuzzy controllers and identifiers for nonlinear dynamic systems and present rigorous stability and performance analyses for them.

In Chapter 2, we first review basic concepts of fuzzy sets and fuzzy logic, and then present a detailed description for each component of the fuzzy logic system. We derive specific formula of a number of useful fuzzy logic systems and show that these fuzzy logic systems are universal approximators. That is, they are capable of approximating any nonlinear function over a compact set to arbitrary accuracy.

In Chapter 3, we show that the fuzzy logic systems can be represented as feedforward networks, and based on this network representation we develop back-propagation training algorithms to adjust the parameters of the fuzzy logic systems. We use the resulting adaptive fuzzy systems as identifiers for nonlinear dynamic systems and compare the performance of these fuzzy identifiers with neural-network-based identifiers.

A weak point of the back-propagation algorithm in Chapter 3 is that it performs a nonlinear search procedure and may converge slowly or be trapped at a local minimum. To overcome this weak point, we develop another training algorithm in Chapter 4 that uses the classical orthogonal least squares algorithm. Specifically, we first fix some parameters in the fuzzy logic systems so that the fuzzy logic systems can be represented as linear combinations of the so-called fuzzy basis functions; then, we use the orthogonal least squares algorithm to select the significant fuzzy basis functions and the corresponding optimal coefficients. We use this adaptive fuzzy system to approximate a controller for the nonlinear ball-and-beam system.

Although the training algorithms in Chapters 3 and 4 have well-justified performance criteria, they are not very simple in the sense that their computational requirements may be intense for complex problems. In Chapter 5, we develop a very simple training algorithm which just performs a one-pass operation on the training data. The basic idea is to generate fuzzy IF-THEN rules based on training data and then combine the generated rules with the rules from human experts into the final fuzzy logic system. We apply this adaptive fuzzy system to the truck backer-upper control and time-series prediction problems.

The training algorithms in Chapters 3–5 cannot guarantee that the trained fuzzy logic system can match all the input-output pairs in the training set to any given accuracy. If there is only a limited number of input-output pairs, we may want a fuzzy logic system which can match all the input-output pairs to any given accuracy. In Chapter 6, we show how to design such an optimal fuzzy logic system. We also develop adaptive versions of the optimal fuzzy logic system based on nearest neighborhood clustering, and apply the resulting adaptive fuzzy systems to controlling nonlinear dynamic systems.

In Chapter 7, we compare the adaptive fuzzy systems with artificial neural networks, including multilayer perceptron, radial basis function network, and probabilistic general regression. We show that the most important advantage of adaptive fuzzy systems is that they are capable of combining numerical information from sensors and linguistic information from human experts into a common framework in a systematic and efficient manner, whereas the artificial neural networks can only make use of numerical information.

In Chapter 8, we first discuss the reasons for fuzzy control and classify adaptive fuzzy controllers into a few categories: indirect versus direct, and first type versus second type. Then, we design two indirect adaptive fuzzy controllers for higher-order nonlinear systems based on the Lyapunov synthesis approach. We prove that: (1) the resulting closed-loop systems with these adaptive fuzzy controllers are globally stable in the sense that all variables involved (states, controls, parameters, and so on) are uniformly bounded (that is, stable in the bounded-input–bounded-output sense), and (2) the tracking error between the system output and the reference trajectory converges to zero under some conditions. We use these adaptive fuzzy controllers to control the inverted pendulum to track a given trajectory. The indirect adaptive fuzzy controllers are capable of incorporating fuzzy descriptions about the unknown system under control into the controllers.

In Chapter 9, we design two direct adaptive fuzzy controllers which are capable of incorporating fuzzy control rules into the controllers. We prove that these direct adaptive fuzzy controllers guarantee a globally stable closed-loop system and that the tracking error converges to zero under some conditions. We use them to control an unstable system and the chaotic Duffing forced-oscillation system.

The adaptive fuzzy controllers in Chapters 8 and 9 are developed for the higher-order nonlinear system in the canonical form: $x^{(n)} = f(\underline{x}) + g(\underline{x})u$, where $\underline{x} = (x, \dot{x}, \ldots, x^{(n-1)})^T$, and $f(\underline{x})$ and $g(\underline{x})$ are unknown nonlinear functions, not for the general nonlinear system: $\underline{\dot{x}} = f(\underline{x}) + g(\underline{x})u$, where $\underline{x}$ is an arbitrary state vector. In Chapter 10, we develop adaptive fuzzy controllers for the general nonlinear system using the basic concepts of input-output linearization in the nonlinear control literature. We first review the basic concepts of input-output linearization, and then use them to design adaptive fuzzy controllers. We simulate the adaptive fuzzy controllers to control the ball-and-beam system to track a trajectory.

Another way of designing adaptive fuzzy controllers for the general nonlinear system is to first estimate the nonlinear system based on adaptive fuzzy systems and then design controllers based on the estimation model. In Chapter 11, we design two identifiers for the general nonlinear system based on adaptive fuzzy systems. We prove that all variables in the fuzzy identifiers are uniformly bounded and that the estimation model converges to the true system under some conditions. We simulate the fuzzy identifiers for the chaotic glycolytic oscillator.

In Chapter 12, we develop two nonlinear adaptive filters based on adaptive fuzzy systems: recursive least squares (RLS) and least mean squares (LMS) fuzzy adaptive filters, and use them as equalizers for nonlinear communication channels. We show that the fuzzy adaptive filters can incorporate fuzzy descriptions about the unknown channels into the equalizers.

Finally, Chapter 13 contains general conclusions of the book and a discussion of future research.

2

DESCRIPTION AND ANALYSIS OF FUZZY LOGIC SYSTEMS

2.1 BASIC CONCEPTS OF FUZZY SETS AND FUZZY LOGIC

In this section, we briefly review basic concepts of fuzzy sets and fuzzy logic which will be useful in describing fuzzy logic systems.

2.1.1 Fuzzy Set and Set-Theoretical Operators

Definition 2.1. *Fuzzy Set*: Let U be a collection of objects, for example, $U = R^n$, and be called the universe of discourse. A fuzzy set F in U is characterized by a membership function $\mu_F : U \to [0, 1]$, with $\mu_F(u)$ representing the grade of membership of $u \in U$ in the fuzzy set F. A fuzzy set may be viewed as a generalization of the concept of an ordinary set (that is, a crisp set) whose membership function only takes two values $\{0, 1\}$.

Figure 2.1 shows the membership functions of three fuzzy sets, namely, "slow," "medium," and "fast" for the speed of a car. In this example, the universe of discourse is all possible speeds of the car; that is, $U = [0, V_{max}]$, where V_{max} is the maximum speed of the car. At a speed of 45 mph, for example, the fuzzy set "slow" has membership value 0.5, that is, $\mu_{slow}(45) = 0.5$, the fuzzy set "medium" has membership value 0.5, that is, $\mu_{medium}(45) = 0.5$, and the fuzzy set "fast" has membership value 0, that is, $\mu_{fast}(45) = 0$.

Definition 2.2. *Support, Center, and Fuzzy Singleton*: The support of a fuzzy set F is the crisp set of all points $u \in U$ such that $\mu_F(u) > 0$. The center of a fuzzy set F is the point(s) $u \in U$ at which $\mu_F(u)$ achieves its maximum value.

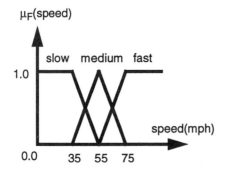

Figure 2.1 Membership functions of three fuzzy sets, namely, "slow," "medium," and "fast" for the speed of a car.

If the support of a fuzzy set F is a single point in U at which $\mu_F = 1$, the F is called a fuzzy singleton.

Definition 2.3. *Intersection, Union, and Complement*: Let A and B be two fuzzy sets in U. The intersection $A \cap B$ of A and B is a fuzzy set in U with membership function defined for all $u \in U$ by

$$\mu_{A \cap B}(u) = min\{\mu_A(u), \mu_B(u)\} \tag{2.1}$$

The union of $A \cup B$ of A and B is a fuzzy set in U with the membership defined for all $u \in U$ by

$$\mu_{A \cup B}(u) = max\{\mu_A(u), \mu_B(u)\} \tag{2.2}$$

Usually, the intersection and union operators are denoted by $\wedge$ and $\vee$, respectively. The complement $\bar{A}$ of A is a fuzzy set in U with the membership function defined for all $u \in U$ by

$$\mu_{\bar{A}}(u) = 1 - \mu_A(u) \tag{2.3}$$

Definition 2.3 shows only one possible choice of operators for intersection, union, and complement. One choice of operator corresponds to one interpretation of the meaning of the logic operation's intersection, union, and complement. Based on different interpretations which range from intuitive argumentation to empirical or axiomatic justifications, other operators have been suggested in the literature. For example, the so-called t-norm and t-conorm were proposed for intersection and union operations, respectively.

Definition 2.4. *T-norm and T-conorm*: A t-norm, denoted by $\star$, is a two-place function from $[0, 1] \times [0, 1]$ to $[0, 1]$, which includes fuzzy intersection, algebraic product, bounded product, and drastic product, defined as

$$x \star y = \begin{cases} min\{x, y\} & fuzzy\ intersection \\ xy & algebraic\ product \\ max\{0, x + y - 1\} & bounded\ product \\ x & if\ y = 1 \\ y & if\ x = 1 \\ 0 & if\ x, y < 1 \end{cases} \quad drastic\ product \tag{2.4}$$

where $x, y \in [0, 1]$. A t-conorm, denoted by $\dot{+}$, is a two-place function from $[0, 1] \times [0, 1]$ to $[0, 1]$, which includes fuzzy union, algebraic sum, bounded sum, and drastic sum, defined as

$$x \dot{+} y = \begin{cases} max\{x, y\} & fuzzy\ union \\ x + y - xy & algebraic\ sum \\ min\{1, x + y\} & bounded\ sum \\ \begin{cases} x & if\ y = 0 \\ y & if\ x = 0 \\ 1 & if\ x, y > 0 \end{cases} & drastic\ sum \end{cases} \quad (2.5)$$

where $x, y \in [0, 1]$. Other types of t-norm and t-conorm are also proposed in the literature, but for the purpose of this book, (2.4) and (2.5) are sufficient.

2.1.2 The Extension Principle

The extension principle is a tool for generalizing crisp mathematical concepts to fuzzy sets. It has been extensively used in the fuzzy literature.

Definition 2.5. *The Extension Principle*: Let U and V be two universes of discourse and f be a mapping from U to V. For a fuzzy set A in U, the extension principle defines a fuzzy set B in V by

$$\mu_B(v) = sup_{u \in f^{-1}(v)}[\mu_A(u)] \quad (2.6)$$

That is, $\mu_B(v)$ is the superium of $\mu_A(u)$ for all $u \in U$ such that $f(u) = v$, where $v \in V$ and we assume that $f^{-1}(v)$ is not empty. If $f^{-1}(v)$ is empty for some $v \in V$, define $\mu_B(v) = 0$.

2.1.3 Fuzzy Relations and Their Compositions

Definition 2.6. *Fuzzy Relation*: Let U and V be two universes of discourse. A fuzzy relation R is a fuzzy set in the product space $U \times V$; that is, R has the membership function $\mu_R(u, v)$, where $u \in U$ and $v \in V$.

Definition 2.7. *Sup-Star Composition*: Let R and S be fuzzy relations in $U \times V$ and $V \times W$, respectively. The sup-star composition of R and S is a fuzzy relation denoted by $R \circ S$ and is defined by

$$\mu_{R \circ S}(u, w) = sup_{v \in V}[\mu_R(u, v) \star \mu_S(v, w)] \quad (2.7)$$

where $u \in U$, $w \in W$, and $\star$ could be any operator in the class of t-norm defined by (2.4). Clearly, $R \circ S$ is a fuzzy set in $U \times W$. It is possible that S is just a fuzzy set in V; in this case, the $\mu_S(v, w)$ in (2.7) becomes $\mu_S(v)$, the $\mu_{R \circ S}(u, w)$ becomes $\mu_{R \circ S}(u)$, and others remain the same.

The most commonly used sup-star compositions are the sup-min and sup-product compositions which replace the ⋆ in (2.7) by min and algebraic product, respectively.

2.1.4 Linguistic Variables and Hedges

There are two interpretations for the concept of a linguistic variable. Formally, a linguistic variable is defined as follows [Zadeh,108]:

Definition 2.8. *Linguistic Variables (formal)*: A linguistic variable is characterized by a quintuple $(x, T(x), U, G, M)$ in which x is the name of variable; $T(x)$ is the term set of x, that is, the set of names of linguistic values of x with each value being a fuzzy set defined on U; G is a syntactic rule for generating the names of values of x; and M is a semantic rule for associating each value with its meaning.

This definition may give the reader a feeling that the linguistic variable is a complex concept, but in fact it should not be. The goal of introducing the concept of a linguistic variable is to present a formal way of saying that a variable may take words in natural languages as its values. For example, if we can say "the speed is fast," then the variable *speed* should be understood as a linguistic variable; but this does not mean that the variable *speed* cannot take real values. In this spirit, we have the following intuitive definition of a linguistic variable.

Definition 2.9. *Linguistic Variables (intuitive)*: If a variable can take words in natural languages (for example, small, fast, and so on) as its values, this variable is defined as a linguistic variable. These words are usually labels of fuzzy sets. A linguistic variable can take either words or numbers as its values.

For example, the linguistic variable *speed* can take "slow," "medium," and "fast" as its values, as defined in Fig. 2.1; it can also take any real numbers in the interval $[0, V_{max}]$ as its values. The linguistic variable is an important concept that gives us a formal way to quantify linguistic descriptions about variables.

Since in linguistic descriptions we often use hedges such as "very" and "more or less" to describe other terms, we need formal definitions for what these hedges mean. Although in everyday use the hedge "very" does not have a well-defined meaning, in essence it acts as an intensifier. In this spirit, we have the following definitions for the two most commonly used hedges: "very" and "more or less."

Definition 2.10. *Hedges*: Let F be a fuzzy set in U (for example, F=small), then "very F" is defined as a fuzzy set in U with the membership function

$$\mu_{very\ F}(u) = (\mu_F(u))^2 \tag{2.8}$$

and "more or less F" is a fuzzy set in U with membership function

$$\mu_{more\ or\ less\ F}(u) = (\mu_F(u))^{1/2} \tag{2.9}$$

where $u \in U$.

2.1.5 Generalized Modus Ponens and Generalized Modus Tollens

In fuzzy logic and approximate reasoning, there are two important fuzzy inference rules, namely, generalized modus ponens (GMP) and generalized modus tollens (GMT).

Definition 2.11. *Generalized Modus Ponens (GMP)*: GMP is defined as the following inference procedure:

premise 1: x is A'

premise 2: IF x is A, THEN y is B

consequence: y is B'

where A', A, B, and B' are fuzzy sets, and x and y are linguistic variables.

Table 2.1 shows intuitive criteria relating premise 1 and the consequence in GMP. We note that if a causal relation between "x is A" and "y is B" is not strong in premise 2, the satisfaction of criterion 2-2 and criterion 3-2 is allowed. Criterion 4-2 is interpreted as: "IF x is A THEN y is B, else y is not B." Although this relation is not valid in formal logic, we often make such an interpretation in everyday reasoning.

TABLE 2.1 Intuitive criteria relating premise 1 and the consequence for given premise 2 in GMP

	x is A' (premise 1)	y is B' (consequence)
criterion 1	x is A	y is B
criterion 2-1	x is very A	y is very B
criterion 2-2	x is very A	y is B
criterion 3-1	x is more or less A	y is more or less B
criterion 3-2	x is more or less A	y is B
criterion 4-1	x is not A	y is unknown
criterion 4-2	x is not A	y is not B

Definition 2.12. *Generalized Modus Tollens (GMT)*: GMT is defined as the following inference procedure:

premise 1: y is B'

premise 2: IF x is A, THEN y is B

consequence: x is A'

where A', A, B', and B are fuzzy sets, and x and y are linguistic variables.

TABLE 2.2 Intuitive criteria relating premise 1 and the consequence for given premise 2 in GMT

	y is B' (premise 1)	x is A' (consequence)
criterion 5	y is not B	x is not A
criterion 6	y is not very B	x is not very A
criterion 7	y is not more or less B	x is not more or less A
criterion 8-1	y is B	x is unknown
criterion 8-2	y is B	x is A

Table 2.2 shows some intuitive criteria relating premise 1 and the consequence in GMT.

2.1.6 Fuzzy Implications

Definition 2.13. *Fuzzy Implication*: Let A and B be fuzzy sets in U and V, respectively. A fuzzy implication, denoted by $A \to B$, is a special kind of fuzzy relation in $U \times V$ with the following membership functions:

- Fuzzy conjunction:

$$\mu_{A \to B}(u, v) = \mu_A(u) \star \mu_B(v) \qquad (2.10)$$

- Fuzzy disjunction:

$$\mu_{A \to B}(u, v) = \mu_A(u) \dotplus \mu_B(v) \qquad (2.11)$$

- Material implication:

$$\mu_{A \to B}(u, v) = \mu_{\bar{A}}(u) \dotplus \mu_B(v) \qquad (2.12)$$

- Propositional calculus:

$$\mu_{A \to B}(u, v) = \mu_{\bar{A}}(u) \dotplus \mu_{A \star B}(v) \qquad (2.13)$$

- Generalization of modus ponens:

$$\mu_{A \to B}(u, v) = sup\{c \in [0, 1] | \mu_A(u) \star c \leq \mu_B(v)\} \qquad (2.14)$$

- Generalization of modus tollens:

$$\mu_{A \to B}(u, v) = inf\{c \in [0, 1] | \mu_B(v) \dotplus c \leq \mu_A(u)\} \qquad (2.15)$$

A fuzzy implication $A \to B$ can be understood as a fuzzy IF-THEN rule: IF x is A, THEN y is B, where $x \in U$ and $y \in V$ are linguistic variables. (2.10)-(2.15) correspond to six interpretations of the fuzzy IF-THEN rule based on intuitive criteria or generalizations of classical logic.

2.2 FUZZY RULE BASE

In Figure 1.3 of Section 1.2 we showed the basic configuration of the fuzzy logic systems considered in this book. In this and the next three sections, we present a detailed description of the four basic components: fuzzy rule base, fuzzy inference engine, fuzzifier, and defuzzifier, respectively.

A *fuzzy rule base* consists of a collection of fuzzy IF-THEN rules in the following form:

$$R^{(l)}: IF\ x_1\ is\ F_1^l\ and\ \cdots\ and\ x_n\ is\ F_n^l,\ THEN\ y\ is\ G^l \quad (2.16)$$

where F_i^l and G^l are fuzzy sets in $U_i \subset R$ and $V \subset R$, respectively, and $\underline{x} = (x_1, \ldots, x_n)^T \in U_1 \times \cdots \times U_n$ and $y \in V$ are linguistic variables. Let M be the number of fuzzy IF-THEN rules in the form of (2.16) in the fuzzy rule base; that is, $l = 1, 2, \ldots, M$ in (2.16). The $\underline{x}$ and y are the input and output to the fuzzy logic system, respectively. Without loss of generality, we consider multi-input–single-output fuzzy logic systems, since a multi-output system can always be decomposed into a group of single-output systems.

The fuzzy rule base is the heart of the fuzzy logic system in the sense that all other three components are used to interpret these rules and make them usable for specific problems. Practice has shown that fuzzy IF-THEN rules in the form of (2.16) provide a very convenient framework for human experts to express their domain knowledge.

We may have the following basic questions concerning the fuzzy rule base:

- Are the rules in the form of (2.16) general enough to include other types of linguistic information?
- Where do these fuzzy IF-THEN rules come from?
- How are the membership functions for F_i^l and G^l determined?

We answer the first question by showing that the rules in the form of (2.16) include many other types of fuzzy rules as special cases.

Fact 2.1. The rules in the form of (2.16) include the following "incomplete-IF-part rule" as a special case:

$$IF\ x_1\ is\ F_1^l\ and\ \cdots\ and\ x_m\ is\ F_m^l,\ THEN\ y\ is\ G^l$$

where $m < n$.

Proof. Clearly, the preceding incomplete-IF-part rule is equivalent to

$$IF\ x_1\ is\ F_1^l\ and\ \cdots\ and\ x_m\ is\ F_m^l\ and\ x_{m+1}\ is\ I1\ and\ \cdots$$
$$and\ x_n\ is\ I1,\ THEN\ y\ is\ G^l$$

where $I1$ is a fuzzy set in R with $\mu_{I1}(x) \equiv 1$ for all $x \in R$. The preceding rule is in the form of (2.16); thus, this fact is true. Q.E.D.

Fact 2.2. The rules in the form of (2.16) include the following "OR rule" as a special case:

IF x_1 is F_1^l and $\cdots$ and x_m is F_m^l or x_{m+1} is F_{m+1}^l and $\cdots$ and x_n is F_n^l, THEN y is G^l

Proof. Based on intuitive meaning of the logic operator "or," the preceding OR rule is equivalent to the following two rules:

IF x_1 is F_1^l and $\cdots$ and x_m is F_m^l, THEN y is G^l

IF x_{m+1} is F_{m+1}^l and $\cdots$ and x_n is F_n^l, THEN y is G^l

From Fact 2.1 we have that the preceding two rules are special cases of (2.16); therefore, this fact is true. Q.E.D.

Fact 2.3. The rules in the form of (2.16) include the fuzzy statement:

$$y \text{ is } G^l$$

as a special case.

Proof. Clearly, the preceding fuzzy statement is equivalent to

IF x_1 is $I1$ and $\cdots$ and x_n is $I1$, THEN y is G^l

which is in the form of (2.16). Q.E.D.

Fact 2.4. The rules in the form of (2.16) include the following "gradual rule" as a special case:

The smaller the x, the bigger the y

Proof. Let S be a fuzzy set representing "smaller," for example, $\mu_S(x) = 1/(1 + exp(5(x+2)))$, and B be a fuzzy set representing "bigger," for example, $\mu_B(y) = 1/(1 + exp(-5(y-2)))$, then the preceding "gradual rule" is equivalent to

IF x is S, THEN y is B

which is a special case of (2.16). Q.E.D.

Fact 2.5. The rules in the form of (2.16) include the following "unless rule" as a special case:

y is G^l unless x_1 is F_1^l and $\cdots$ and x_n is F_n^l

Proof. Based on the intuitive meaning of *unless*, the preceding rule is equivalent to

IF not (x_1 is F_1^l and $\cdots$ and x_n is F_n^l), THEN y is G^l

which, based on the De Morgan's Law, is equivalent to

IF x_1 is not F_1^l or $\cdots$ or x_n is not F_n^l, THEN y is G^l

View *not* F_i^l as a single fuzzy set, then from Fact 2.2 the preceding rule is a special case of (2.16). Q.E.D.

Fact 2.6. The rules in the form of (2.16) include non-fuzzy rules (i.e., conventional production rules) as a special case.

Proof. If the membership functions of F_i^l and G^l can only take values 1 or 0, then the rules (2.16) become non-fuzzy rules. Q.E.D.

For the second question, there are two principal ways of obtaining fuzzy IF-THEN rules: (1) asking human experts, and (2) using training algorithms based on measured data. The first way is the most straightforward method for obtaining rules, but human experts in many cases may not provide a sufficient number of rules. The second way is the main focus of this book; details are given in the following chapters.

For the third question, there are also two principal ways depending upon where the rules come from. If the rules are provided by human experts, then the membership functions of F_i^l and G^l should be specified by the experts because these functions are an integrated part of the expert knowledge. For example, if an expert says that: "IF the error is large, THEN the control is large," he or she should tell what the *large* means by specifying the fuzzy membership functions for the *large*. If the rules are determined by numerical data, then the first task is to determine the functional forms for $\mu_{F_i^l}$ and μ_{G^l}. The most commonly used functional forms are Gaussian, triangular, and trapezoid. Gaussian functions have a smooth transition property, whereas triangular and trapezoid functions are simpler to compute. After the functional forms of $\mu_{F_i^l}$ and μ_{G^l} are fixed, the problem becomes how to determine the parameters in $\mu_{F_i^l}$ and μ_{G^l} based on measured data. We develop a number of parameter estimation methods for $\mu_{F_i^l}$ and μ_{G^l} in later chapters.

2.3 FUZZY INFERENCE ENGINE

In a *fuzzy inference engine*, fuzzy logic principles are used to combine the fuzzy IF-THEN rules in the fuzzy rule base into a mapping from fuzzy sets in $U = U_1 \times \cdots \times U_n$ to fuzzy sets in V. The first question for the fuzzy inference engine is: How do we interpret a fuzzy IF-THEN rule in the form of (2.16)?

2.3.1 Interpretations of a Fuzzy IF-THEN Rule

A fuzzy IF-THEN rule (2.16) is interpreted as a fuzzy implication $F_1^l \times \cdots \times F_n^l \to G^l$ in $U \times V$. Let a fuzzy set A' in U be the input to the fuzzy inference engine; then each fuzzy IF-THEN rule (2.16) determines a fuzzy set B^l in V using the sup-star composition (2.7). That is,

$$\mu_{B^l}(y) = sup_{\underline{x} \in U}[\mu_{F_1^l \times \cdots \times F_n^l \to G^l}(\underline{x}, y) \star \mu_{A'}(\underline{x})] \qquad (2.17)$$

From Definition 2.13 we see that there are six interpretations (2.10)–(2.15) for a fuzzy implication, and in each interpretation we may employ different t-norms or t-conorms; therefore, a fuzzy IF-THEN rule (2.16) can be interpreted in a number of ways. In the following, we show some commonly used interpretations for the fuzzy IF-THEN rule. For simplicity, we denote $F_1^l \times \cdots \times F_n^l = A$ and $G^l = B$, and the rule (2.16) is therefore denoted by $A \to B$.

- Mini-operation rule of fuzzy implication:

$$\mu_{A \to B}(\underline{x}, y) = min\{\mu_A(\underline{x}), \mu_B(y)\} \tag{2.18}$$

- Product-operation rule of fuzzy implication:

$$\mu_{A \to B}(\underline{x}, y) = \mu_A(\underline{x})\mu_B(y) \tag{2.19}$$

- Arithmetic rule of fuzzy implication:

$$\mu_{A \to B}(\underline{x}, y) = min\{1, 1 - \mu_A(\underline{x}) + \mu_B(y)\} \tag{2.20}$$

- Maxmin rule of fuzzy implication:

$$\mu_{A \to B}(\underline{x}, y) = max\{min[\mu_A(\underline{x}), \mu_B(y)], 1 - \mu_A(\underline{x})\} \tag{2.21}$$

- Boolean rule of fuzzy implication:

$$\mu_{A \to B}(\underline{x}, y) = max\{1 - \mu_A(\underline{x}), \mu_B(y)\} \tag{2.22}$$

- Goguen's rule of fuzzy implication:

$$\mu_{A \to B}(\underline{x}, y) = \begin{cases} 1 & \mu_A(\underline{x}) \leq \mu_B(y) \\ \mu_B(y)/\mu_A(\underline{x}) & \mu_A(\underline{x}) > \mu_B(y) \end{cases} \tag{2.23}$$

In (2.18)–(2.23), $\mu_A(\underline{x}) = \mu_{F_1^l \times \cdots \times F_n^l}(\underline{x})$ is defined either according to the mini-operation rule:

$$\mu_{F_1^l \times \cdots \times F_n^l}(\underline{x}) = min\{\mu_{F_1^l}(x_1), \ldots, \mu_{F_n^l}(x_n)\} \tag{2.24}$$

or according to the product-operation rule:

$$\mu_{F_1^l \times \cdots \times F_n^l}(\underline{x}) = \mu_{F_1^l}(x_1) \cdots \mu_{F_n^l}(x_n) \tag{2.25}$$

We note that the mini-operation rule follows from the fuzzy conjunction (2.10) by using the fuzzy intersection operator in (2.4) for $\star$; the product-operation rule follows from the fuzzy conjunction by using the algebraic product for $\star$. The arithmetic rule follows from the material implication (2.12) by using the bounded sum for $\dot{+}$ ($\mu_{A \to B} = \mu_{\bar{A} \dot{+} B} = min\{1, 1 - \mu_A + \mu_B\}$); the maxmin rule follows from the propositional calculus (2.13) by using the fuzzy intersection for $\star$ and fuzzy union for $\dot{+}$ ($\mu_{A \to B} = \mu_{\bar{A} \dot{+} (A \star B)} = max\{1 - \mu_A, min[\mu_A, \mu_B]\}$). Boolean rule follows from the material implication by using the fuzzy union for $\dot{+}$ ($\mu_{A \to B} = \mu_{\bar{A} \dot{+} B} = max\{1 - \mu_A, \mu_B\}$), and Goguen's rule follows from the generalization of

Sec. 2.3 Fuzzy Inference Engine

modus ponens by using the algebraic product for $\star$ ($\mu_{A \to B} = sup\{c \in [0,1] | \mu_A \cdot c \leq \mu_B\} = 1$ if $\mu_A \leq \mu_B$ and $= \mu_B/\mu_A$ if $\mu_A > \mu_B$).

From (2.18)–(2.25) we see that there are many different ways of interpreting a fuzzy IF-THEN rule. A natural question is: What are the properties of these interpretations?

2.3.2 Properties of the Interpretations

Here we investigate the consequences when we apply (2.18)–(2.23) in the fuzzy inference and, in particular, the GMP and GMT. In the GMP, we examine the consequences B' of the following compositional equation:

$$B' = A' \circ R \qquad (2.26)$$

for the following particular choices of A': $A' = A$, $A' = very\ A$ ((2.8)), $A' = more\ or\ less\ A$ ((2.9)), and $A = \bar{A}$, where R denotes the $A \to B$ in (2.18)–(2.23), and $\circ$ is the sup-min operator, that is, (2.7) with $\star = min$. Similarly, in the GMT, we examine the consequences A' of the following equation:

$$A' = R \circ B' \qquad (2.27)$$

for $B' = \bar{B}$, *not very B*, *not more or less B*, and B.

First, we examine the product-operation rule (2.19) in GMP. In the following we assume that $sup_{x \in U}[\mu_A(x)] = 1$. If $A' = A$, we have:

$$\mu_{B'}(y) = sup_{x \in U}\{min[\mu_A(x)\mu_B(y), \mu_A(x)]\}$$
$$= sup_{x \in U}[\mu_A(x)\mu_B(y)]$$
$$= \mu_B(y) \qquad (2.28)$$

If $A' = very\ A$, we have:

$$\mu_{B'}(y) = sup_{x \in U}\{min[\mu_A(x)\mu_B(y), \mu_A^2(x)]\}$$
$$= \mu_B(y) \qquad (2.29)$$

If $A' = more\ or\ less\ A$, we have:

$$\mu_{B'}(y) = sup_{x \in U}\{min[\mu_A(x)\mu_B(y), \mu_A^{1/2}(x)]\}$$
$$= \mu_B(y) \qquad (2.30)$$

And, if $A' = \bar{A}$, we have:

$$\mu_{B'}(y) = sup_{x \in U}\{min[\mu_A(x)\mu_B(y), 1 - \mu_A(x)]\}$$
$$= \frac{\mu_B(y)}{1 + \mu_B(y)} \qquad (2.31)$$

Similarly, we can determine the consequences for other interpretations in (2.18)–(2.23). Table 2.3 summarizes the consequences $\mu_{B'}$ for the four particular instances of A' with the six interpretations (2.18)–(2.23).

TABLE 2.3 Summary of inference results for GMP[a]

	A	very A	more or less A	$\bar{A}$
R_c	μ_B	μ_B	μ_B	$0.5 \wedge \mu_B$
R_p	μ_B	μ_B	μ_B	$\frac{\mu_B}{1+\mu_B}$
R_a	$\frac{1+\mu_B}{2}$	$\frac{3+2\mu_B-\sqrt{5+4\mu_B}}{2}$	$\frac{\sqrt{5+4\mu_B}-1}{2}$	1
R_m	$0.5 \vee \mu_B$	$\frac{3-\sqrt{5}}{2} \vee \mu_B$	$\frac{\sqrt{5}-1}{2} \vee \mu_B$	1
R_b	$0.5 \vee \mu_B$	$\frac{3-\sqrt{5}}{2} \vee \mu_B$	$\frac{\sqrt{5}-1}{2} \vee \mu_B$	1
R_g	$\sqrt{\mu_B}$	$(\mu_B)^{2/3}$	$(\mu_B)^{1/3}$	1

[a]Where R_c refers to mini-operation rule (2.18), R_p refers to product-operation rule (2.19), R_a refers to arithmetic rule (2.20), R_m refers to maxmin rule (2.21), R_b refers to Boolean rule (2.22), and R_g refers to Goguen's rule (2.23).

Next, we examine the product-operation rule (2.19) in GMT. We assume that $sup_{y \in V}[\mu_B(y)] = 1$. If $B' = \bar{B}$, we have:

$$\mu_{A'}(\underline{x}) = sup_{y \in V}\{min[\mu_A(\underline{x})\mu_B(y), 1 - \mu_B(y)]\}$$
$$= \frac{\mu_A(\underline{x})}{1 + \mu_A(\underline{x})} \tag{2.32}$$

If $B' = not\ very\ B$, we have:

$$\mu_{A'}(\underline{x}) = sup_{y \in V}\{min[\mu_A(\underline{x})\mu_B(y), 1 - \mu_B^2(y)]\}$$
$$= \frac{\mu_A(\underline{x})\sqrt{\mu_A^2(\underline{x}) + 4} - \mu_A^2(\underline{x})}{2} \tag{2.33}$$

If $B' = not\ more\ or\ less\ B$, we have:

$$\mu_{A'}(\underline{x}) = sup_{y \in V}\{min[\mu_A(\underline{x})\mu_B(y), 1 - \mu_B^{1/2}(y)]\}$$
$$= \frac{1 + 2\mu_A(\underline{x}) - \sqrt{1 + 4\mu_A(\underline{x})}}{2\mu_A(\underline{x})} \tag{2.34}$$

And, if $B' = B$, we have:

$$\mu_{A'}(\underline{x}) = sup_{y \in V}\{min[\mu_A(\underline{x})\mu_B(y), \mu_B(y)]\}$$
$$= \mu_A(\underline{x}) \tag{2.35}$$

Similarly, we can determine the consequences for other interpretations in (2.18)–(2.23), which are shown in Table 2.4.

By employing the intuitive criteria in Tables 2.1 and 2.2 in Tables 2.3 and 2.4, we can determine how well an interpretation satisfies them. This information is summarized in Table 2.5. From Table 2.5 we see that the mini-operation rule

Sec. 2.3 Fuzzy Inference Engine

TABLE 2.4 Summary of inference results for GMT

	$\bar{B}$	not very B	not more or less B	B
R_c	$0.5 \wedge \mu_A$	$\frac{\sqrt{5}-1}{2} \wedge \mu_A$	$\frac{3-\sqrt{5}}{2} \wedge \mu_A$	μ_A
R_p	$\frac{\mu_A}{1+\mu_A}$	$\frac{\mu_A\sqrt{\mu_A^2+4}-\mu_A^2}{2}$	$\frac{2\mu_A+1-\sqrt{4\mu_A+1}}{2\mu_A}$	μ_A
R_a	$1-\frac{\mu_A}{2}$	$\frac{1-2\mu_A+\sqrt{1+4\mu_A}}{2}$	$\frac{3-\sqrt{1+\mu_A}}{2}$	1
R_m	$0.5 \vee (1-\mu_A)$	$(1-\mu_A) \vee \left(\frac{\sqrt{5}-1}{2} \wedge \mu_A\right)$	$\frac{3-\sqrt{5}}{2} \vee (1-\mu_A)$	$\mu_A \vee (1-\mu_A)$
R_b	$0.5 \vee (1-\mu_A)$	$\frac{\sqrt{5}-1}{2} \vee (1-\mu_A)$	$\frac{3-\sqrt{5}}{2} \vee (1-\mu_A)$	1
R_g	$\frac{1}{1+\mu_A}$	$\frac{\sqrt{1+4\mu_A^2}-1}{2\mu_A^2}$	$\frac{2+\mu_A-\sqrt{\mu_A^2+4\mu_A}}{2}$	1

TABLE 2.5 Satisfaction of various fuzzy inference rules [a]

	R_c	R_p	R_a	R_m	R_b	R_g
criterion 1	Y	Y	N	N	N	N
criterion 2-1	N	N	N	N	N	N
criterion 2-2	Y	Y	N	N	N	N
criterion 3-1	N	N	N	N	N	N
criterion 3-2	Y	Y	N	N	N	N
criterion 4-1	N	N	Y	Y	Y	Y
criterion 4-2	N	N	N	N	N	N
criterion 5	N	N	N	N	N	N
criterion 6	N	N	N	N	N	N
criterion 7	N	N	N	N	N	N
criterion 8-1	N	N	Y	N	Y	Y
criterion 8-2	Y	Y	N	N	N	N

[a] Under intuitive criteria of Tables 2.1 and 2.2, where "Y" represents yes and "N" represents no.

and the product-operation rule are good inference rules because they fit many of the intuitive criteria.

2.3.3 Overall Mapping of the Fuzzy Inference Engine

For an input A' (a fuzzy set in U), the output of the fuzzy inference engine can take two forms: (1) M fuzzy sets $B^l (l = 1, 2, \ldots, M)$ in the form of (2.17), with each one determined by one fuzzy IF-THEN rule in the form of (2.16), and (2) one fuzzy set B' which is the union of the M fuzzy sets B^l. That is,

$$\mu_{B'}(y) = \mu_{B^1}(y) \dot{+} \cdots \dot{+} \mu_{B^M}(y) \tag{2.36}$$

That is, for a single input, the output of the fuzzy inference engine can either be a collection of M fuzzy sets or be the union of the M fuzzy sets. For these different types of outputs, we use different defuzzifiers to defuzzify them into a single point in the output space V.

2.4 FUZZIFIER

The *fuzzifier* performs a mapping from a crisp point $\underline{x} = (x_1, \ldots, x_n)^T \in U$ into a fuzzy set A' in U. There are (at least) two possible choices of this mapping:

- *Singleton fuzzifier*: A' is a fuzzy singleton with support $\underline{x}$, that is, $\mu_{A'}(\underline{x}') = 1$ for $\underline{x}' = \underline{x}$ and $\mu_{A'}(\underline{x}') = 0$ for all other $\underline{x}' \in U$ with $\underline{x}' \neq \underline{x}$.
- *Nonsingleton fuzzifier*: $\mu_{A'}(\underline{x}) = 1$ and $\mu_{A'}(\underline{x}')$ decreases from 1 as $\underline{x}'$ moves away from $\underline{x}$, for example, $\mu_{A'}(\underline{x}') = exp\left[-\frac{(\underline{x}'-\underline{x})^T(\underline{x}'-\underline{x})}{\sigma^2}\right]$, where σ^2 is a parameter characterizing the shape of $\mu_{A'}(\underline{x}')$.

It seems that only the singleton fuzzifier has been used. We think that the nonsingleton fuzzifier may be useful if the inputs are corrupted by noise. This issue needs further studying.

2.5 DEFUZZIFIER

The *defuzzifier* performs a mapping from fuzzy sets in V to a crisp point $y \in V$. There are (at least) three possible choices of this mapping:

- *Maximum defuzzifier*, defined as

$$y = argsup_{y \in V}(\mu_{B'}(y)) \qquad (2.37)$$

where $\mu_{B'}(y)$ is given by (2.36).
- *Center average defuzzifier*, defined as

$$y = \frac{\sum_{l=1}^{M} \bar{y}^l(\mu_{B^l}(\bar{y}^l))}{\sum_{l=1}^{M}(\mu_{B^l}(\bar{y}^l))} \qquad (2.38)$$

where $\bar{y}^l$ is the center of the fuzzy set G^l, that is, the point in V at which $\mu_{G^l}(y)$ achieves its maximum value, and $\mu_{B^l}(y)$ is given by (2.17).
- *Modified center average defuzzifier*, defined as

$$y = \frac{\sum_{l=1}^{M} \bar{y}^l(\mu_{B^l}(\bar{y}^l)/\delta^l)}{\sum_{l=1}^{M}(\mu_{B^l}(\bar{y}^l)/\delta^l)} \qquad (2.39)$$

where δ^l is a parameter characterizing the shape of $\mu_{G^l}(y)$ such that the narrower the shape of $\mu_{G^l}(y)$, the smaller is δ^l; for example, if $\mu_{G^l}(y) = exp[-(\frac{y-\bar{y}^l}{\delta^l})^2]$, then δ^l is such a parameter.

The modified center average defuzzifier is justified as follows. Common sense indicates that the sharper the shape of $\mu_{G^l}(y)$, the stronger is our belief that the output y should be nearer to the center of G^l (according to the rule $R^{(l)}$ of (2.16)). The standard center average defuzzifier, (2.38), is a weighted average of the $\bar{y}^l$'s, and the weights $\mu_{B^l}(\bar{y}^l)$ determined by (2.17) do not take the shape of $\mu_{G^l}(y)$ into consideration. This is clearly not satisfactory based on our common sense. An obvious improvement is the modified center average defuzzifier (2.39).

2.6 USEFUL CLASSES OF FUZZY LOGIC SYSTEMS

2.6.1 How to Select Appropriate Operators in the Fuzzy Logic Systems

From Section 2.3 we see that there are many different interpretations for the fuzzy IF-THEN rules which result in different mappings of the fuzzy inference engine. Also, in Sections 2.4 and 2.5 we see that we have different types of fuzzifiers and defuzzifiers. Many combinations of these fuzzy inference engines, fuzzifiers, and defuzzifiers may constitute useful fuzzy logic systems. Therefore, one may ask:

- How do we choose the inference rules (among (2.18)–(2.23)) in the fuzzy inference engine?
- How is the fuzzifier chosen?
- How is the defuzzifier chosen?
- How do we determine the functional forms for the membership functions of the fuzzy sets F_i^l and G^l in the fuzzy IF-THEN rules?

In general, we have the following criteria for these questions:

- *Empirical Fit*: Because fuzzy logic systems are used to incorporate linguistic information, it is important that the choices generate appropriate models of real-system behavior.
- *Axiomatic Strength*: We may view the intuitive criteria in Table 2.1 as possible axioms, and a selection of fuzzy logic principles is better if it satisfies more axioms.
- *Computational Efficiency*: For large problems or limited computing power, we may have to select simpler fuzzy logic systems.
- *Easy for Adaptation*: Because we will develop training algorithms for the fuzzy logic systems, the selections must result in systems which are easy to adapt. For example, if a selection generates a fuzzy logic system which is linear in its adjustable parameters, it is a good choice according to this criterion.

For the fuzzy inference engine, we see in subsection 2.3.2 (in particular Table 2.5) that the mini-inference rule and the product-inference rule are good choices from an axiomatic strength point of view. Also, these inference rules are computationally simple. We will use these inference rules in our adaptive fuzzy systems.

For the two proposed fuzzifiers, the singleton fuzzifier will produce simple fuzzy logic systems (see the next subsection), whereas the nonsingleton fuzzifier may work better in a noisy environment. In this book we do not consider noise-corrupted systems; therefore, we use only the singleton fuzzifier.

For the first two defuzzifiers in Section 2.5, practical experiments have shown that the center average defuzzifier outperforms the maximum defuzzifier. Therefore, we do not consider the maximum defuzzifier in this book. For the center average and modified center average defuzzifiers, the former produces simpler systems and therefore faster training algorithms, and the latter is more powerful in quantifying linguistic information. We will consider both.

For the choice of membership functions of F_i^l and G^l, Gaussian, triangular, and trapezoid are the most commonly used types. We use Gaussian and triangular membership functions in our adaptive fuzzy systems.

In summary, we will consider fuzzy logic systems which are different combinations of the mini-inference rule, the product-inference rule, singleton fuzzifier, center average defuzzifier, modified center average defuzzifier, Gaussian membership function, and triangular membership function. The detailed functional forms of these fuzzy logic systems are derived next.

2.6.2 Derivation of Useful Classes of Fuzzy Logic Systems

Lemma 2.1. The fuzzy logic systems with center average defuzzifier (2.38), product-inference rule (2.19) and (2.25), and singleton fuzzifier are of the following form:

$$f(\underline{x}) = \frac{\sum_{l=1}^{M} \bar{y}^l \left(\prod_{i=1}^{n} \mu_{F_i^l}(x_i) \right)}{\sum_{l=1}^{M} \left(\prod_{i=1}^{n} \mu_{F_i^l}(x_i) \right)} \quad (2.40)$$

where $\bar{y}^l$ is the point at which μ_{G^l} achieves its maximum value, and we assume that $\mu_{G^l}(\bar{y}^l) = 1$.

Proof. Using the center average defuzzifier (2.38), we have

$$f(\underline{x}) = \frac{\sum_{l=1}^{M} \bar{y}^l (\mu_{B^l}(\bar{y}^l))}{\sum_{l=1}^{M} (\mu_{B^l}(\bar{y}^l))} \quad (2.41)$$

where $\mu_{B^l}(\bar{y}^l)$ is given by (2.17). Using the product-inference rule (2.19) and (2.25) in (2.17), we have

$$\mu_{B^l}(\bar{y}^l) = sup_{\underline{x}' \in U} \left[\prod_{i=1}^{n} \mu_{F_i^l}(x_i') \mu_{G^l}(\bar{y}^l) \mu_{A'}(\underline{x}') \right] \quad (2.42)$$

Sec. 2.6 Useful Classes of Fuzzy Logic Systems

If we use the singleton fuzzifier, we have $\mu_{A'}(\underline{x}') = 1$ for $\underline{x}' = \underline{x}$ ($\underline{x}$ is the input crisp point to the fuzzy logic system) and $\mu_{A'}(\underline{x}') = 0$ for all other $\underline{x}' \in U$; therefore, the sup in (2.42) is achieved at $\underline{x}' = \underline{x}$, and (2.42) can be simplified to

$$\mu_{B^l}(\bar{y}^l) = \prod_{i=1}^{n} \mu_{F_i^l}(x_i) \tag{2.43}$$

where we assume that $\mu_{G^l}(\bar{y}^l) = 1$. Substituting (2.43) into (2.41), we obtain (2.40). Q.E.D.

Similarly, we have the following lemma.

Lemma 2.2. The fuzzy logic systems with center average defuzzifier (2.38), mini-inference rule (2.18) and (2.24), and singleton fuzzifier are of the following form:

$$f(\underline{x}) = \frac{\sum_{l=1}^{M} \bar{y}^l [min(\mu_{F_1^l}(x_1), \ldots, \mu_{F_n^l}(x_n))]}{\sum_{l=1}^{M} [min(\mu_{F_1^l}(x_1), \ldots, \mu_{F_n^l}(x_n))]} \tag{2.44}$$

where $\bar{y}^l$ is the point at which μ_{G^l} achieves its maximum value, and we assume that $\mu_{G^l}(\bar{y}^l) = 1$.

Proof. Using the same method as in the proof of Lemma 2.1 and replacing the $\prod$ by min, we can prove this lemma. Q.E.D.

In order to develop training algorithms for these fuzzy logic systems, we need to specify the functional form for $\mu_{F_i^l}$. Our first choice is the following Gaussian function:

$$\mu_{F_i^l}(x_i) = a_i^l exp\left[-\left(\frac{x_i - \bar{x}_i^l}{\sigma_i^l}\right)^2\right] \tag{2.45}$$

where $a_i^l, \bar{x}_i^l$, and σ_i^l are adjustable parameters.

Lemma 2.3. The fuzzy logic systems with center average defuzzifier (2.38), product-inference rule (2.19) and (2.25), singleton fuzzifier, and Gaussian membership function (2.45) are of the following form:

$$f(\underline{x}) = \frac{\sum_{l=1}^{M} \bar{y}^l \left[\prod_{i=1}^{n} a_i^l exp\left(-\left(\frac{x_i - \bar{x}_i^l}{\sigma_i^l}\right)^2\right)\right]}{\sum_{l=1}^{M} \left[\prod_{i=1}^{n} a_i^l exp\left(-\left(\frac{x_i - \bar{x}_i^l}{\sigma_i^l}\right)^2\right)\right]} \tag{2.46}$$

Proof. Just substitute (2.45) into (2.40). Q.E.D.

Remark 2.1. The fuzzy logic system (2.46) is the most frequently used fuzzy logic system in this book. The adjustable parameters of this fuzzy logic system are $\bar{y}^l, a_i^l, \bar{x}_i^l$ and σ_i^l, with the constraints $\bar{y}^l \in V$, $a_i^l \in (0, 1)$, $\bar{x}_i^l \in U_i$, and $\sigma_i^l > 0$. We assume that $a_i^l = 1$ because intuitively we can assume that $\mu_{F_i^l}$ achieves 1 at some point. In the following chapters of this book, we develop a variety of training algorithms for adjusting these parameters and apply the resulting adaptive fuzzy systems, that is, the fuzzy logic system (2.46) equipped with the training algorithms, to a variety of control and signal processing problems.

Similarly, we can also use the following triangular membership functions:

$$\mu_{F_i^l}(x_i) = 1 - \frac{|x_i - c_i^l|}{b_i^l} \quad \text{if} \quad x_i \in [c_i^l - b_i^l, c_i^l + b_i^l] \quad (2.47)$$

$$0 \quad otherwise$$

where c_i^l and b_i^l are adjustable parameters.

Finally, we show the functional form of the fuzzy logic systems with the modified center average defuzzifier.

Lemma 2.4. Let $\mu_{F_i^l}$ and μ_{G^l} be Gaussian functions; that is, $\mu_{F_i^l}$ is given by (2.45) and μ_{G^l} is defined by

$$\mu_{G^l}(y) = exp\left[-\left(\frac{y - \bar{y}^l}{\delta^l}\right)^2\right] \quad (2.48)$$

Then the fuzzy logic systems with modified center average defuzzifier (2.39), product-inference rule (2.19) and (2.25), singleton fuzzifier, and Gaussian membership function are of the following form:

$$f(\underline{x}) = \frac{\sum_{l=1}^{M} \bar{y}^l \left[\prod_{i=1}^{n} a_i^l exp\left(-\left(\frac{x_i - \bar{x}_i^l}{\sigma_i^l}\right)^2\right)\right] / \delta^l}{\sum_{l=1}^{M} \left[\prod_{i=1}^{n} a_i^l exp\left(-\left(\frac{x_i - \bar{x}_i^l}{\sigma_i^l}\right)^2\right)\right] / \delta^l} \quad (2.49)$$

Proof. Use (2.39) instead of (2.38) and then use the same procedure as the proof of Lemma 2.1. Q.E.D.

There are two fundamental questions concerning these classes of fuzzy logic systems:

- How is linguistic information incorporated using these fuzzy logic systems?
- What is the capability of these fuzzy logic systems as nonlinear mappings from a functional approximation point of view?

The first question is easy to answer. Because these fuzzy logic systems are constructed from the fuzzy IF-THEN rules (2.16), linguistic information in the form of (2.16) can be directly incorporated. We point out this question because we want to remind the reader that (2.40), (2.44), (2.46), and (2.49) are not just nonlinear functions from $\underline{x} \in U$ to $f(\underline{x}) \in V$, they are *fuzzy logic systems with particular choices of fuzzy inference engines, fuzzifiers, defuzzifiers, and fuzzy membership functions.*

The second question is important because the prime use of these fuzzy logic systems is as models of nonlinear systems, including human operators. Therefore, if these fuzzy logic systems are capable of approximating a wide variety of nonlinear functions, they are qualified as models of general nonlinear systems; otherwise, it is difficult to justify their use as nonlinear system models. We answer this question next.

2.7 FUZZY LOGIC SYSTEMS AS UNIVERSAL APPROXIMATORS

The following theorem shows that the fuzzy logic systems in Lemma 2.3 are capable of uniformly approximating any nonlinear function over U to any degree of accuracy if U is compact.

Universal Approximation Theorem. For any given real continuous function g on a compact set $U \subset R^n$ and arbitrary $\epsilon > 0$, there exists a fuzzy logic system f in the form of (2.46) such that

$$sup_{\underline{x} \in U} |f(\underline{x}) - g(\underline{x})| < \epsilon \qquad (2.50)$$

Proof of this theorem is given in the Appendix.
We can extend this result to discrete functions, as follows.

Corollary. For any $g \in L_2(U)$ and arbitrary $\epsilon > 0$, there exists a fuzzy logic system f in the form of (2.46) such that

$$\left(\int_U |f(\underline{x}) - g(\underline{x})|^2 d\underline{x} \right)^{1/2} < \epsilon \qquad (2.51)$$

where $U \subset R^n$ is compact, $L_2(U) = [g : U \to R | \int_U |g(\underline{x})|^2 d\underline{x} < \infty]$, and the integrals are in the Lebesgue sense.

Proof of this corollary is also given in the Appendix.
We now make a few remarks on this Universal Approximation Theorem.

Remark 2.2. This theorem provides a justification for applying the fuzzy logic systems to almost any nonlinear modeling problems. It also provides an explanation for the practical successes of the fuzzy logic systems in engineering applications.

Remark 2.3. Since the fuzzy logic system in the form of (2.46) is a special case of the fuzzy logic systems in the form of (2.40), the latter are also universal approximators.

Remark 2.4. This theorem is just an existence theorem; that is, it shows that there *exists* a fuzzy logic system (2.46) that can uniformly approximate any given function to arbitrary accuracy. How to find such a fuzzy logic system is another question. Although we use this theorem as one justification for using the fuzzy logic systems, the importance of this theorem should not be overemphasized because many other types of functions are also universal approximators, including the simple polynormals [Rudin, 60]. What should be emphasized is the capability of the fuzzy logic systems to incorporate linguistic information in a natural and systematic way—a unique advantage of the fuzzy logic systems which is not shared by other types of universal approximators, including polynormals, neural networks, and so on. This issue is further discussed in Chapter 7.

2.8 CONCLUDING REMARKS

From this chapter we see that the fuzzy logic systems comprise a very rich collection of nonlinear functions due to the availability of a variety of fuzzy logic principles. By specifying the fuzzy logic principles used in the fuzzy logic systems and other factors like fuzzifier and defuzzifier, we obtain specific formula of some particular fuzzy logic systems. We show that even these particular fuzzy logic systems are general enough to approximate any nonlinear function to arbitrary accuracy.

In the following chapters, we concentrate on some particular fuzzy logic systems and develop various training algorithms to adjust their parameters based on numerical data. We use the resulting adaptive fuzzy systems to solve engineering problems and analyze their performance.

3

TRAINING OF FUZZY LOGIC SYSTEMS USING BACK-PROPAGATION

3.1 INTRODUCTION

After a relatively silent period in the 1970s, the research on artificial neural networks has gained strong public interest during the recent years. One reason for this resurgent interest is the discovery of a powerful training algorithm for multilayer neural networks—the so-called back-propagation algorithm (see Section 7.2 for a brief review of this algorithm). In fact, the basic concept of the back-propagation algorithm can be applied to any feedforward networks. Therefore, if we can represent the fuzzy logic systems as feedforward networks, we can use the idea of back-propagation to train them. This is the motivation of the training algorithm in this chapter.

Since we showed in Section 2.7 that the fuzzy logic systems in the form of (2.46) are universal approximators, we develop a back-propagation training algorithm for these kinds of fuzzy logic systems in this chapter. By observing the functional form of (2.46), we see that it can be represented as a three-layer feedforward network shown in Figure 3.1. With this network representation of the fuzzy logic systems, it becomes straightforward to apply the back-propagation idea to adjust the parameters $\bar{y}^l$, $\bar{x}_i^l$, and σ_i^l, that is, to train the fuzzy logic system. The derivation of this back-propagation algorithm is given next.

3.2 BACK-PROPAGATION TRAINING ALGORITHM FOR THE FUZZY LOGIC SYSTEMS

Suppose that we are given an input-output pair $(\underline{x}^p, d^p)$, $\underline{x}^p \in U \subset R^n$, $d^p \in V \subset R$; our task is to determine a fuzzy logic system $f(\underline{x})$ in the form of (2.46) such

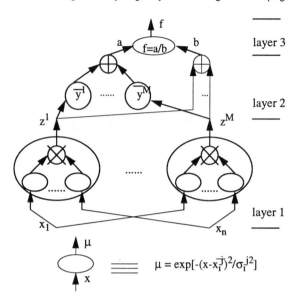

Figure 3.1 Network representation of the fuzzy logic systems.

that

$$e^p = \frac{1}{2}[f(\underline{x}^p) - d^p]^2 \tag{3.1}$$

is minimized. We assume that $a_i^l = 1$ and M is given; therefore, the problem becomes training the parameters $\bar{y}^l, \bar{x}_i^l$, and σ_i^l such that e^p of (3.1) is minimized. In the following we use e, f, and d to denote e^p, $f(\underline{x}^p)$, and d^p, respectively.

To train $\bar{y}^l$ we use

$$\bar{y}^l(k+1) = \bar{y}^l(k) - \alpha \frac{\partial e}{\partial \bar{y}^l}|_k \tag{3.2}$$

where $l = 1, 2, \ldots, M$, $k = 0, 1, 2, \ldots$, and α is a constant stepsize. From Figure 3.1 we see that f (and hence e) depends on $\bar{y}^l$ only through a, where $f = a/b$, $a = \sum_{l=1}^{M}(\bar{y}^l z^l)$, $b = \sum_{l=1}^{M} z^l$, and $z^l = \prod_{i=1}^{n} exp(-(\frac{x_i - \bar{x}_i^l}{\sigma_i^l})^2)$; hence, using the chain rule, we have

$$\frac{\partial e}{\partial \bar{y}^l} = (f - d)\frac{\partial f}{\partial a}\frac{\partial a}{\partial \bar{y}^l} = (f - d)\frac{1}{b}z^l \tag{3.3}$$

Substituting (3.3) into (3.2), we obtain the training algorithm for $\bar{y}^l$:

$$\bar{y}^l(k+1) = \bar{y}^l(k) - \alpha \frac{f - d}{b} z^l \tag{3.4}$$

where $l = 1, 2, \ldots, M$, and $k = 0, 1, 2, \ldots$.

Sec. 3.2 Back-Propagation Training Algorithm for the Fuzzy Logic Systems

To train $\bar{x}_i^l$, we use

$$\bar{x}_i^l(k+1) = \bar{x}_i^l(k) - \alpha \frac{\partial e}{\partial \bar{x}_i^l}\bigg|_k \tag{3.5}$$

where $i = 1, 2, \ldots, n, l = 1, 2, \ldots, M$, and $k = 0, 1, 2, \ldots$. We see from Figure 3.1 that f (and hence e) depends on $\bar{x}_i^l$ only through z^l; hence, using the chain rule, we have

$$\frac{\partial e}{\partial \bar{x}_i^l} = (f-d)\frac{\partial f}{\partial z^l}\frac{\partial z^l}{\partial \bar{x}_i^l} = (f-d)\frac{\bar{y}^l - f}{b}z^l\frac{2(x_i^p - \bar{x}_i^l)}{\sigma_i^{l2}} \tag{3.6}$$

Substituting (3.6) into (3.5), we obtain the training algorithm for $\bar{x}_i^l$:

$$\bar{x}_i^l(k+1) = \bar{x}_i^l(k) - \alpha \frac{f-d}{b}(\bar{y}^l - f)z^l\frac{2(x_i^p - \bar{x}_i^l(k))}{\sigma_i^{l2}(k)} \tag{3.7}$$

where $i = 1, 2, \ldots, n, l = 1, 2, \ldots, M$, and $k = 0, 1, 2, \ldots$.

Using the same method as previously, we obtain the following training algorithm for σ_i^l:

$$\sigma_i^l(k+1) = \sigma_i^l(k) - \alpha \frac{\partial e}{\partial \sigma_i^l}\bigg|_k$$

$$= \sigma_i^l(k) - \alpha \frac{f-d}{b}(\bar{y}^l - f)z^l\frac{2(x_i^p - \bar{x}_i^l(k))^2}{\sigma_i^{l3}(k)} \tag{3.8}$$

where $i = 1, 2, \ldots, n, l = 1, 2, \ldots, M$, and $k = 0, 1, 2, \ldots$.

The training algorithm (3.4), (3.7), and (3.8) performs an error back-propagation procedure. To train $\bar{y}^l$, the "normalized" error $(f-d)/b$ is back-propagated to the layer of $\bar{y}^l$; then $\bar{y}^l$ is updated using (3.4) in which z^l is the input to $\bar{y}^l$ (see Figure 3.1). To train $\bar{x}_i^l$ and σ_i^l, the "normalized" error $(f-d)/b$ times $(\bar{y}^l - f)$ and z^l is back-propagated to the processing unit of Layer 1 whose output is z^l; then $\bar{x}_i^l$ and σ_i^l are updated using (3.7) and (3.8), respectively, in which the remaining variables $\bar{x}_i^l, x_i^p$, and σ_i^l (that is, the variables on the right-hand sides of (3.7) and (3.8), except the back-propagated error $\frac{f-d}{b}(\bar{y}^l - f)z^l$) can be obtained locally.

The training procedure for the fuzzy system of Figure 3.1 is a two-pass procedure. First, for a given input $\underline{x}^p$, compute forward along the network (that is, the fuzzy logic system) to obtain z^l ($l = 1, 2, \ldots, M$), a, b, and f. Then train the network parameters $\bar{y}^l, \bar{x}_i^l$, and σ_i^l ($i = 1, 2, \ldots, n, l = 1, 2, \ldots, M$) backwards using (3.4), (3.7), and (3.8), respectively. For multiple input-output pairs, that is, we are given $(\underline{x}^p, d^p)$ with $p = 1, 2, \ldots$, we may train the system for one or more forward-backward cycles for an input-output pair before moving to the next pair.

3.3 APPLICATION TO NONLINEAR DYNAMIC SYSTEM IDENTIFICATION

3.3.1 Motivation

In their award-winning paper, Narendra and Parthasarathy [49] used the back-propagation neural networks as identifiers for nonlinear components in dynamic systems. Theoretical justification of this approach is that feedforward neural networks can approximate any real continuous function on a compact set to arbitrary accuracy [Cybenko, 12; Hornik, Stinchcombe, and White, 20]. The back-propagation training algorithm [Rumelhart and McCleland, 61; Werbos, 95] makes it possible to train the neural network identifiers on line to match unknown nonlinear mappings. Through a large number of simulations, they showed that the neural identifiers can accurately identify a variety of nonlinear dynamic systems.

Because we showed in Section 2.7 that the fuzzy logic systems are also universal approximators and we also have a back-propagation algorithm to train the fuzzy logic systems, we can use the fuzzy logic systems equipped with the back-propagation training algorithm as identifiers for nonlinear dynamic systems. We call this kind of identifiers *fuzzy identifiers*, and the identifiers of Narendra and Parthasarathy [49] *neural identifiers*. Now, before applying these fuzzy identifiers to the examples in Narendra and Parthasarathy [49], one may ask: "Why use the fuzzy identifiers instead of using the neural identifiers?" or "What are the conceptual advantages of the fuzzy identifiers over the neural identifiers?"

3.3.2 Conceptual Advantages of the Fuzzy Identifiers Over the Neural Identifiers

There are two advantages of the fuzzy identifiers over the neural identifiers:

- The parameters of the fuzzy identifiers have clear physical meanings; that is, the $\bar{y}^l$'s are the centers of the THEN part fuzzy sets of the fuzzy IF-THEN rules, and the $\bar{x}_i^l$'s and σ_i^l's are the centers and widths of the IF part fuzzy sets; hence, as we show later in this section, it is possible to develop a very good method for choosing the initial parameters. On the other hand, the parameters of the neural identifiers have no clear relationships with input-output data, and therefore their initial values are usually chosen randomly. Because both back-propagation training algorithms are gradient algorithms, good initial parameters dramatically speed convergence.

- Because the fuzzy identifiers are constructed based on fuzzy logic systems which are constructed from a set of fuzzy IF-THEN rules, the fuzzy identifiers provide a natural framework in which to incorporate human linguistic descriptions (in the form of IF-THEN rules) about the unknown nonlinear system. Specifically, we use the linguistic information to help to build a good initial identifier, so that the identifier will converge to the true system faster during the on-line training. This capability of the fuzzy identifiers has practical importance because many real-world nonlinear systems are controlled by human experts (for example, aircraft, power systems, economic systems, and so on), and these experts can provide linguistic descriptions about the nonlinear systems. These linguistic descriptions are vague and fuzzy, so that traditional identifiers [Goodwin and Payne, 16; Ljung, 38] and neural identifiers [Narendra and Parthasarathy, 49] cannot make use of them at the front end of their designs. They are used only to help evaluate such designs. Using the fuzzy identifiers, the identifier designer will be able to include the linguistic information at the front end of the design, where we believe it will do the most good.

3.3.3 Design of the Fuzzy Identifiers

Consider the discrete nonlinear system
$$y(k+1) = f(y(k), \ldots, y(k-n+1); u(k), \ldots, u(k-m+1)) \quad (3.9)$$
where f is an unknown function we want to identify, u and y are the input and output of the system, respectively, and n and m are positive integers. Our task is to identify the unknown function f based on fuzzy logic systems.

As shown in Narendra and Parthasarathy [49], there are two schemes for identifying (3.9):

- Parallel model:
$$\hat{y}(k+1) = \hat{f}(\hat{y}(k), \ldots, \hat{y}(k-n+1); u(k), \ldots, u(k-m+1)) \quad (3.10)$$
where $\hat{f}$ is a fuzzy logic system, and $\hat{y}$ is the output of the identification model, that is, the $\hat{f}$.
- Series-parallel model:
$$\hat{y}(k+1) = \hat{f}(y(k), \ldots, y(k-n+1); u(k), \ldots, u(k-m+1)) \quad (3.11)$$
where y is the output of the system (3.9).

That is, in the parallel model the output of the identification model is fed back to the model, while in the series-parallel model the output of the system is fed to the model. Figures 3.2 and 3.3 show the parallel and series-parallel models, respectively. As shown in Narendra and Parthasarathy [49], the series-parallel model is better than the parallel model; therefore, we use the series-parallel model in our fuzzy identifiers.

Sometimes the unknown function f is a combination of a known linear part and an unknown nonlinear part [Narendra and Parthasarathy, 49]. In this case, we only model the unknown nonlinear part by the fuzzy logic system. Figure 3.4 shows such an example, where α_0 and α_1 are known.

The design of the fuzzy identifier consists of two parts: (1) initial fuzzy logic system $\hat{f}$ construction, and (2) on-line adaptation of $\hat{f}$. The initial $\hat{f}$ should be

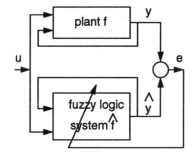

Figure 3.2 Basic scheme of the parallel identification model using fuzzy logic systems.

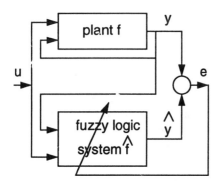

Figure 3.3 Basic scheme of series-parallel identification model using fuzzy logic systems.

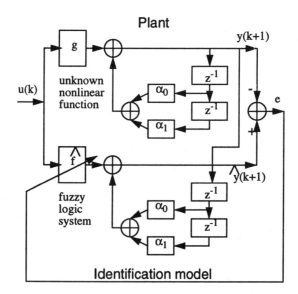

Figure 3.4 An example of series-parallel identification model using fuzzy logic systems.

Sec. 3.3 Application to Nonlinear Dynamic System Identification

constructed to the best approximation of the unknown function f based on the initially available information, while the on-line adaptation adjusts the parameters of $\hat{f}$ such that the error e between the system output y and the identification model output $\hat{y}$ is minimized. We use our back-propagation training algorithm (3.4), (3.7), and (3.8) for the on-line adaptation. Now the question is how to determine a good initial fuzzy logic system $\hat{f}$ in the form of (2.46), that is, how to determine the initial parameters $\bar{y}^l$, $\bar{x}_i^l$, and σ_i^l.

We propose the following method for on-line initial parameter-choosing and provide a theoretical justification (Theorem 3.1) for why this is a good method. In order to simplify the notation, the method is proposed for the particular system configuration of Figure 3.4; extensions to other system configurations are straightforward.

- *An On-Line Initial Parameter-Choosing Method for the Fuzzy Identifier*: Suppose the nonlinear system to be identified starts operation from $k = 0$. Do not start the back-propagation training algorithm (3.4), (3.7), and (3.8) for the first M time points. Set the parameters $\bar{x}_i^l(M) = u_i(l)$ and $\bar{y}^l(M) = g(\underline{u}(l))$, where $i = 1, 2, \ldots, n$, $l = 1, 2, \ldots, M$, $\underline{u}(l) = (u_1(l), \ldots, u_n(l))^T$ is the input to both the system and the identification model, and $g(\underline{u}(l))$ is the desired output of the fuzzy logic system $\hat{f}$ for input $\underline{u}(l)$; and set $\sigma_i^l(M)$ equal to some small numbers (see Theorem 3.1), or set $\sigma_i^l(M) = [max(u_i(l) : l = 1, 2, \ldots, M) - min(u_i(l) : l = 1, 2, \ldots, M)]/2M$ (this choice makes the input membership functions "uniformly" cover the range of $u_i(l)$ from $l = 1$ to $l = M$; in all the simulations in the next subsection, we use this choice), where $l = 1, 2, \ldots, M$, $i = 1, 2, \ldots, n$. Start the on-line training for the fuzzy identifier from time point $M + 1$.

We now show that by choosing the σ_i^l sufficiently small, the fuzzy system with the preceding initial parameters can match all the M input-output pairs $(\underline{u}(l), g(\underline{u}(l)))$, $l = 1, 2, \ldots, M$, to arbitrary accuracy.

Theorem 3.1. For arbitrary $\epsilon > 0$, there exists $\sigma^* > 0$ such that the fuzzy logic system $\hat{f}$ (2.46) with the preceding initial $\bar{x}_i^l$ and $\bar{y}^l$ and $\sigma_i^l = \sigma^*$ has the property that

$$|\hat{f}(\underline{u}(l)) - g(\underline{u}(l))| < \epsilon \tag{3.12}$$

for all $l = 1, 2, \ldots, M$.

Proof. From the initial parameter-choosing procedure and (2.46) (with $a_i^l = 1$), we have that the fuzzy logic system with the initial parameters $\bar{x}_i^l(M)$ and $\bar{y}^l(M)$ and $\sigma_i^l = \sigma^*$ is

$$\hat{f}(\underline{u}(l)) = \frac{\sum_{j=1}^{M} g(\underline{u}(j))(\prod_{i=1}^{n} exp[-(u_i(l) - u_i(j))^2/\sigma^{*2}])}{\sum_{j=1}^{M}(\prod_{i=1}^{n} exp[-(u_i(l) - u_i(j))^2/\sigma^{*2}])}$$

$$= \frac{g(\underline{u}(l)) + \sum_{j \neq l=1}^{M} g(\underline{u}(j))(\prod_{i=1}^{n} exp[-(u_i(l) - u_i(j))^2/\sigma^{*2}])}{1 + \sum_{j \neq l=1}^{M}(\prod_{i=1}^{n} exp[-(u_i(l) - u_i(j))^2/\sigma^{*2}])} \quad (3.13)$$

where $l = 1, 2, \ldots, M$. First, assume that $\underline{u}(l) \neq \underline{u}(j)$ for $l \neq j$; thus, there exists some i such that $u_i(l) \neq u_i(j)$; hence, for arbitrary $\epsilon_1 > 0$ and any $l, j = 1, 2, \ldots, M$, $l \neq j$, we can make $\prod_{i=1}^{n} exp[-(u_i(l) - u_i(j))^2/\sigma^{*2}] < \epsilon_1$ by properly choosing σ^* because $exp[-(u_i(l) - u_i(j))^2/\sigma^{*2}] \to 0$ as $\sigma^* \to 0$ if $u_i(l) \neq u_i(j)$. From this result and (3.13) we conclude that there exists $\sigma^* > 0$ such that $|\hat{f}(\underline{u}(l)) - g(\underline{u}(l))| < \epsilon$ for all $l = 1, 2, \ldots, M$.

If $\underline{u}(l) = \underline{u}(j_0)$ for some $j_0 \neq l$, and there are $r - 1$ such j_0; then (3.13) can be written as

$$\hat{f}(\underline{u}(l)) = \frac{rg(\underline{u}(l)) + \sum_j g(\underline{u}(j))(\prod_{i=1}^{n} exp[-(u_i(l) - u_i(j))^2/\sigma^{*2}])}{r + \sum_j (\prod_{i=1}^{n} exp[-(u_i(l) - u_i(j))^2/\sigma^{*2}])} \quad (3.14)$$

where the $\sum_j$ is over all j's in $[1, 2, \ldots, M]$ except l and the j_0's. Using the same arguments as earlier, we can prove the truth of (3.12). Q.E.D.

Based on Theorem 3.1 we can say that the initial parameter-choosing method is a good one because the fuzzy logic system with these initial parameters can at least match the first M input-output pairs arbitrarily well. If these first M input-output pairs contain some important features of the unknown nonlinear mapping, we may hope that after the back-propagation training starts from time point $M + 1$, the fuzzy identifier will converge to the unknown nonlinear system very quickly. In fact, based on our simulation results in the next subsection, this is indeed true. However, we cannot choose σ_i^l to be too small because, although a fuzzy system with small σ_i^l matches the first M pairs quite well, it will have large approximation errors for other input-output pairs. Therefore, in our simulations, we use the second choice of σ_i^l described in the on-line initial parameter-choosing method.

Now suppose that there are L fuzzy IF-THEN rules describing the unknown function f:

$$R_f^j: \; IF \; x_1 \; is \; F_1^j \; and \; \cdots \; and \; x_n \; is \; F_n^j, \; THEN \; f(\underline{x}) \; is \; G^j \quad (3.15)$$

where F_i^j and G^j are fuzzy sets, and $j = 1, 2, \ldots, L$. We assume that $\mu_{F_i^j}$ are Gaussian functions (2.45) with $a_i^l = 1$. We construct the initial $\hat{f}$ in the following way:

- **An On-Line Initial Parameter-Choosing Method with Linguistic Information**: Suppose the unknown system starts operation from $k = 0$. Do not start the back-propagation training algorithm (3.4), (3.7), and (3.8) for the first $M - L$ time points (we assume that $M \geq L$). Set the parameters $\bar{x}_i^l(M - L)$ and $\sigma_i^l(M - L)$ equal to the centers and widths of F_i^l, respectively, and $\bar{y}^l$ equal to the centers of G^l, where $l = 1, 2, \ldots, L$ and $i = 1, 2, \ldots, n$. Set $\bar{x}_i^{l+L}(M - L) = u_i(l)$ and $\bar{y}^{l+L}(M - L) = g(\underline{u}(l))$ for $l = 1, 2, \ldots, M - L$, and $\sigma_i^{l+L}(M - L) = [max(u_i(l) : l = 1, 2, \ldots, M - L) - min(u_i(l) :$

$l = 1, 2, \ldots, M - L)]/2(M - L)$. Start the on-line back-propagation training from time point $M - L + 1$.

We see that in this initial parameter-choosing method, the initial $\hat{f}$ is constructed from two sets of information: (1) L fuzzy IF-THEN rules in the form of (3.15), and (2) the first $M - L$ input-output pairs $(\underline{u}(l), g(\underline{u}(l)))$, $l = 1, 2, \ldots, M - L$. If there are a sufficient number of fuzzy IF-THEN rules, we may choose $M = L$; in this case, the initial $\hat{f}$ is constructed entirely from the linguistic information. If the linguistic rules (3.15) provide good descriptions about the unknown function f, then the initial $\hat{f}$ so constructed will be close to the f; as a result, the training procedure will converge faster.

3.3.4 Simulations

We used the same examples as in Narendra and Parthasarathy [49] to simulate our fuzzy identifier because we want to compare the fuzzy identifiers with the neural identifiers. We simulated the fuzzy identifier for four examples, with each example emphasizing a specific point. Example 3.1 emphasizes the detailed procedure of how the fuzzy identifier learns to match the unknown nonlinear function as training progresses. Example 3.2 shows how performance is improved by incorporating linguistic rules. Example 3.3 shows how the identifier works when only the initial parameters are used. Finally, Example 3.4 shows how the fuzzy identifier works for a multi-input–multi-output system. We chose $M = 40$ for all four examples. The fuzzy logic system (2.46) with $M = 40$ has $40 \times 3 = 120$ adjustable parameters (corresponding to each rule there are three adjustable parameters: $\bar{y}^l$, $\bar{x}_i^l$, and σ_i^l). In Narendra and Parthasarathy [49] the neural network identifiers had two hidden layers with 10 and 20 neurons in each layer respectively; hence, these neural identifiers had $10 \times 20 = 200$ adjustable parameters. Consequently, from a system complexity point of view (in the sense of number of free parameters), the fuzzy identifiers used for the next four examples are simpler than the neural identifiers used in Narendra and Parthasarathy [49] for the same examples.

Example 3.1. The plant to be identified is governed by the difference equation

$$y(k + 1) = 0.3y(k) + 0.6y(k - 1) + g[u(k)] \quad (3.16)$$

where the unknown function has the form $g(u) = 0.6sin(\pi u) + 0.3sin(3\pi u) + 0.1sin(5\pi u)$. In order to identify the plant, a series-parallel model governed by the difference equation

$$\hat{y}(k + 1) = 0.3y(k) + 0.6y(k - 1) + \hat{f}[u(k)] \quad (3.17)$$

was used, where $\hat{f}[*]$ is of the form (2.46) with $M = 40$ and $a_i^l = 1$. We chose $\alpha = 0.5$ in the back-propagation training algorithm (3.4), (3.7), and (3.8), and we used the first on-line initial parameter-choosing method in the last subsection. We started the training from time point $k = 40$, and trained the parameters $\bar{y}^l$, $\bar{x}_i^l$, and

σ_i^l for one cycle at each time point. That is, we used (3.4), (3.7), and (3.8) once at each time point (in this case the k in (3.4), (3.7), and (3.8) agrees with the k in (3.16) and (3.17)). Figures 3.5–3.7 show the outputs of the plant (solid line) and the identification model (dashed line) when the training was stopped at $k = 200$, 300, and 400, respectively, where the input $u(k) = sin(2\pi k/250)$. We see from Figures 3.5–3.7 that: (1) the output of the identification model follows the output of the plant almost immediately, and still does so when the training was stopped at $k = 200$, 300, and 400; and (2) the identification model approximates the plant more and more accurately as more and more training is performed. In Narendra and Parthasarathy [49] the same plant was identified using a neural network identifier which failed to follow the plant when the training was stopped at $k = 500$; Figure 3.8 shows this result. We see from Figure 3.5 that our fuzzy identifier follows the plant without large errors even when the training was stopped as early as $k = 200$. We think that the main reason for the superior performance of the fuzzy identifier is that we have a very good initial parameter-choosing method. We further test the initial parameter-choosing method in Example 3.3.

If we accept the initial parameter-choosing method as a good one, how about the back-propagation training algorithm itself? Can the latter make the identification model converge to the plant when the initial parameters are chosen randomly? Figure 3.9 shows the outputs of the identification model and the plant for the input $u(k) = sin(2\pi k/250)$ for $1 \leq k \leq 250$ and $501 \leq k \leq 700$ and $u(k) = 0.5sin(2\pi k/250) + 0.5sin(2\pi k/25)$ for $251 \leq k \leq 500$ after the identification model was trained for 5,000 time steps using a random input whose amplitude was uniformly distributed over the interval [-1, 1], where the initial $\bar{y}^l(0), \bar{x}_i^l(0)$, and $\sigma_i^l(0)$ for the training phase were random and uniformly dis-

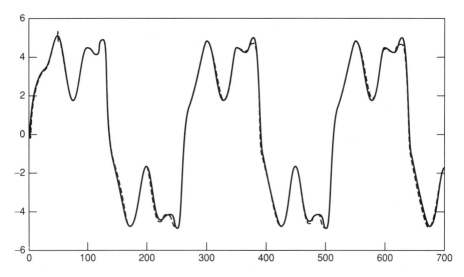

Figure 3.5 Outputs of the plant (solid line) and the identification model (dashed line) for Example 3.1 when the training stops at $k = 200$.

Sec. 3.3 Application to Nonlinear Dynamic System Identification

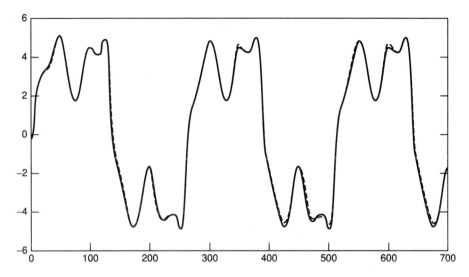

Figure 3.6 Outputs of the plant (solid line) and the identification model (dashed line) for Example 3.1 when the training stops at $k = 300$.

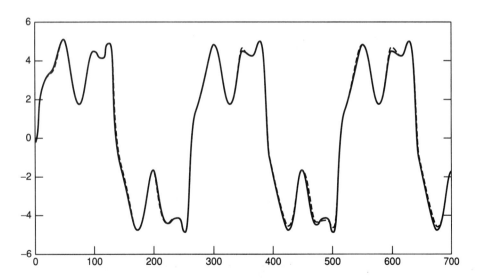

Figure 3.7 Outputs of the plant (solid line) and the identification model (dashed line) for Example 3.1 when the training stops at $k = 400$.

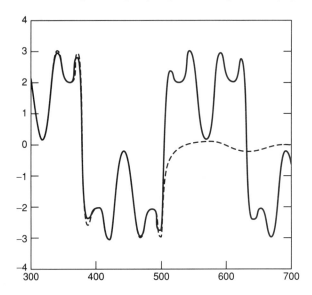

Figure 3.8 Outputs of the plant (solid line) and the neural network identification model of Narendra and Parthasarathy [49, p. 15] (dashed line) for Example 3.1 when the training stops at $k = 500$. (Narendra et al., 1990. © 1990 IEEE.)

tributed over $[-5, 5]$, $[-1, 1]$, and $[0, 0.3]$, respectively. We performed 50 Monte Carlo simulations for the training phase, and all the trained fuzzy identifiers show indistinguishable responses from those shown in Figure 3.9. We see from Figure 3.9 that the trained identification model approximates the plant quite well.

Example 3.2. The plant to be identified is described by the second-order difference equation

$$y(k+1) = g[y(k), y(k-1)] + u(k) \tag{3.18}$$

where

$$g[y(k), y(k-1)] = \frac{y(k)y(k-1)[y(k) + 2.5]}{1 + y^2(k) + y^2(k-1)} \tag{3.19}$$

and $u(k) = sin(2\pi k/25)$. A series-parallel identifier described by the equation

$$\hat{y}(k+1) = \hat{f}[y(k), y(k-1)] + u(k) \tag{3.20}$$

was used, where $\hat{f}[y(k), y(k-1)]$ is in the form of (2.46) with $M = 40$ and $a_i^l = 1$. We chose $\alpha = 0.5$ in the back-propagation training algorithm. Figure 3.10 shows the outputs of the plant and the identification model for arbitrary $\bar{y}^l(0)$, $\bar{x}_i^l(0)$, and $\sigma_i^l(0)$, which were chosen from the intervals [-2,2], [-1,1], and [0, 0.3], respectively, where the fuzzy identifier was trained for one cycle at each time point starting from $k = 0$. Now suppose that we have the following linguistic rule describing

Sec. 3.3 Application to Nonlinear Dynamic System Identification

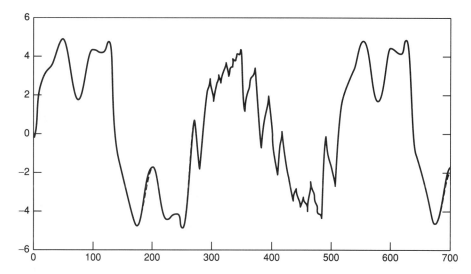

Figure 3.9 Outputs of the identification model (dashed line) and the plant (solid line) for Example 3.1 for the input $u(k) = sin(2\pi k/250)$ for $1 \leq k \leq 250$ and $501 \leq k \leq 700$ and $u(k) = 0.5 sin(2\pi k/250) + 0.5 sin(2\pi k/25)$ for $251 \leq k \leq 500$ after the identification model was trained for 5,000 time steps.

the unknown nonlinear function $g[y(k), y(k-1)]$ of (3.19):

R_g^1 :   IF $y(k)$ is near zero, THEN $g[y(k), y(k-1)]$ is near zero   (3.21)

R_g^2 :   IF $y(k-1)$ is near zero, THEN $g[y(k), y(k-1)]$ is near zero   (3.22)

R_g^3 :   IF $y(k)$ is near -2.5, THEN $g[y(k), y(k-1)]$ is near zero   (3.23)

where "y is near zero" (y can be $y(k)$, $y(k-1)$ or $g[y(k), y(k-1)]$) is characterized by the Gaussian membership function $exp[-\frac{1}{2}(\frac{y}{0.3})^2]$, and "$y$ is near -2.5" is characterized by the Gaussian membership function $exp[-\frac{1}{2}(\frac{y+2.5}{0.3})^2]$. Figure 3.11 shows the outputs of the plant and the identification model after the linguistic rules (3.21)–(3.23) are incorporated, where some initial $\bar{y}^l(0)$, $\bar{x}_i^l(0)$, and $\sigma_i^l(0)$ were determined based on the fuzzy IF-THEN rules (3.21)–(3.23) using the method similar to the second initial parameter-choosing method in the last subsection. The remaining $\bar{y}^l(0)$, $\bar{x}_i^l(0)$, and $\sigma_i^l(0)$ were determined using the same method as in the simulation of Figure 3.10, and the identifier was trained for one cycle at each time point starting from $k = 0$. Comparing Figures 3.11 and 3.10 we see an improvement in adaptation speed after the linguistic rules were incorporated.

Example 3.3. The plant to be identified is of the form

$$y(k+1) = g[y(k), y(k-1), y(k-2), u(k), u(k-1)] \qquad (3.24)$$

where the unknown function g has the form

$$g(x_1, x_2, x_3, x_4, x_5) = \frac{x_1 x_2 x_3 x_5 (x_3 - 1) + x_4}{1 + x_3^2 + x_2^2} \qquad (3.25)$$

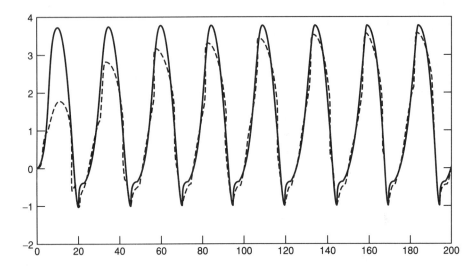

Figure 3.10 Outputs of the plant (solid line) and the identification model (dashed line) for Example 3.2 without using the linguistic rule.

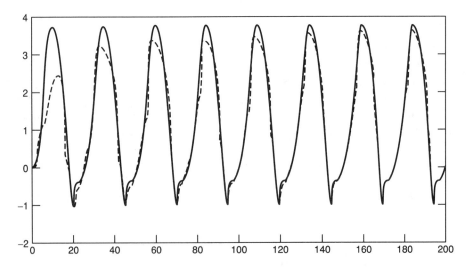

Figure 3.11 Outputs of the plant (solid line) and the identification model (dashed line) for Example 3.2 after the linguistic rule was incorporated.

Sec. 3.3 Application to Nonlinear Dynamic System Identification

and $u(k) = sin(2\pi k/250)$ for $k \leq 500$ and $u(k) = 0.8 sin(2\pi k/250) + 0.2 sin(2\pi k/25)$ for $k > 500$. The identification model is

$$\hat{y}(k+1) = \hat{f}[y(k), y(k-1), y(k-2), u(k), u(k-1)] \quad (3.26)$$

where $\hat{f}$ is of the form (2.46) with $M = 40$ and $a_i^l = 1$. One purpose for this example is to test the first initial parameter-choosing method described in the last subsection. For this purpose, we used the on-line initial parameter-choosing method, but after $k = M$ no back-propagation training was performed. Figure 3.12 shows the outputs of the plant and the identification model whose parameters were determined only based on the on-line initial parameter-choosing method. We see from Figure 3.12 that the identification model could track the plant but with large error. Next, we trained the fuzzy identifier for 5,000 time steps using a random input $u(k)$ whose magnitude was uniformly distributed over [-1,1], where the parameters were trained for one cycle at each time point, and we used the on-line initial parameter-choosing method. We chose $\alpha = 0.5$ in the training phase. The outputs of the plant and the trained fuzzy identifier are shown in Figure 3.13. In Narendra and Parthasarathy [49], a neural network identifier was used for this plant; Figure 3.14 shows the outputs of the plant and the neural identifier after the neural identifier was trained for 10^5 steps using a random input uniformly distributed in the interval $[-1, 1]$. Comparing Figures 3.13 and 3.14, we see that although the neural identifier was trained for 100,000 steps, its performance is worse than that of our fuzzy identifier, which was trained for only 5,000 steps.

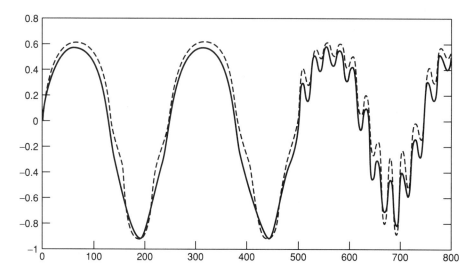

Figure 3.12 Outputs of the plant (solid line) and the identification model (dashed line) for Example 3.3 when the parameters of the identifier were determined based only on the initial parameter-choosing method and there was no back-propagation training.

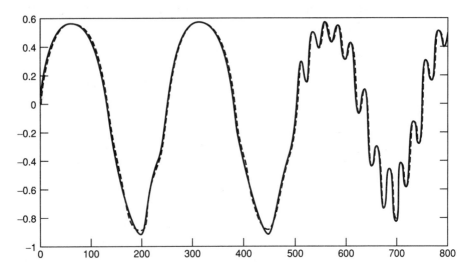

Figure 3.13 Outputs of the plant (solid line) and the identification model (dashed line) for Example 3.3 after 5,000 steps of training.

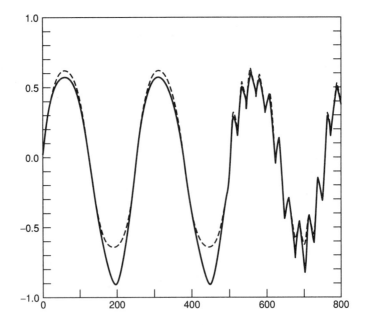

Figure 3.14 Outputs of the plant (solid line) and the neural network identification model of Narendra and Parthasarathy [49, p.17] (dashed-line) for Example 3.3 after 10^5 steps of training. (Narendra et al., 1990. © 1990 IEEE.)

Sec. 3.3 Application to Nonlinear Dynamic System Identification

Example 3.4. In this example, we show how the fuzzy identifier works for a multi-input–multi-output plant which is described by the equations

$$\begin{bmatrix} y_1(k+1) \\ y_2(k+1) \end{bmatrix} = \begin{bmatrix} \frac{y_1(k)}{1+y_2^2(k)} \\ \frac{y_1(k)y_2(k)}{1+y_2^2(k)} \end{bmatrix} + \begin{bmatrix} u_1(k) \\ u_2(k) \end{bmatrix} \quad (3.27)$$

The series-parallel identification model consists of two fuzzy logic systems, $\hat{f}^1$ and $\hat{f}^2$, and is described by the equations

$$\begin{bmatrix} \hat{y}_1(k+1) \\ \hat{y}_2(k+1) \end{bmatrix} = \begin{bmatrix} \hat{f}^1(y_1(k), y_2(k)) \\ \hat{f}^2(y_1(k), y_2(k)) \end{bmatrix} + \begin{bmatrix} u_1(k) \\ u_2(k) \end{bmatrix} \quad (3.28)$$

Both $\hat{f}^1$ and $\hat{f}^2$ are in the form of (2.46) with $M = 40$ and $a_i^l = 1$. The identification procedure was carried out for 5,000 time steps using random inputs $u_1(k)$ and $u_2(k)$ whose magnitudes were uniformly distributed over [-1,1], where we chose $\alpha = 0.5$, used the first on-line initial parameter-choosing method proposed in the last subsection, and trained the parameters for one cycle at each time point. The responses of the plant and the trained identification model for a vector input $[u_1(k), u_2(k)] = [sin(2\pi k/25), cos(2\pi k/25)]$ are shown in Figures 3.15 and 3.16 for $y_1(k)$ and $\hat{y}_1(k)$ and $y_2(k)$ and $\hat{y}_2(k)$, respectively. In Narendra and Parthasarathy [49], a neural network identifier was used for this plant; Figures 3.17 and 3.18 show the outputs of the plant and the neural identifier after the neural identifier was trained for 10^5 steps with inputs u_1 and u_2 uniformly distributed in $[-1, 1]$. Comparing Figures 3.15 and 3.16 with Figures 3.17 and 3.18, we see that the performance of the fuzzy and neural identifiers is similar, although the fuzzy identifier was trained for only 5,000 steps, whereas the neural identifier was trained for 10^5 steps.

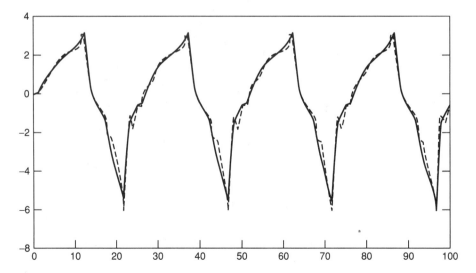

Figure 3.15 Outputs of the plant ($y_1(k)$, solid line) and the identification model ($\hat{y}_1(k)$, dashed line) for Example 3.4 after 5,000 steps of training.

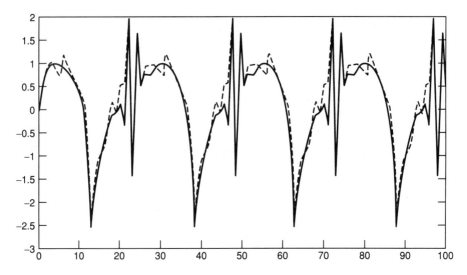

Figure 3.16 Outputs of the plant ($y_2(k)$, solid line) and the identification model ($\hat{y}_2(k)$, dashed line) for Example 3.4 after 5,000 steps of training.

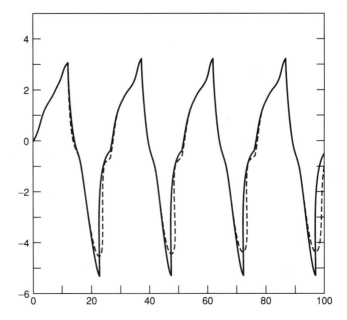

Figure 3.17 Outputs of the plant ($y_1(k)$, solid line) and the neural network identification model of Narendra and Parthasarathy [49, p.18] ($\hat{y}_1(k)$, dashed line) for Example 3.4 after 10^5 steps of training. (Narendra et al., 1990. © 1990 IEEE.)

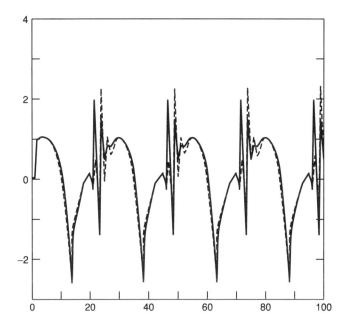

Figure 3.18 Outputs of the plant ($y_2(k)$, solid line) and the neural network identification model of Narendra and Parthasarathy [49, p.18] ($\hat{y}_2(k)$, dashed line) for Example 3.4 after 10^5 steps of training. (Narendra et al., 1990. © 1990 IEEE.)

3.4 CONCLUDING REMARKS

By showing that the fuzzy logic systems can be represented as three-layer feedforward networks, we develop a back-propagation training algorithm for them. The fuzzy logic system equipped with this back-propagation training algorithm constitutes the first adaptive fuzzy system in this book. As compared with the back-propagation neural network, this adaptive fuzzy system has two advantages. (1) The parameters of the adaptive fuzzy system have clear physical meanings, based on which we were able to develop a very good initial parameter-choosing method which greatly speeds up the convergence of the training procedure. On the other hand, the parameters of the back-propagation neural network have no clear physical meanings. Therefore, their initial values have to be chosen randomly, which results in slow convergence. (2) The adaptive fuzzy system can incorporate the linguistic information (in the form of fuzzy IF-THEN rules) in a systematic manner, whereas the back-propagation neural network cannot make use of the linguistic information.

We used the adaptive fuzzy systems as identifiers for nonlinear dynamic systems and simulated the fuzzy identifiers for the same examples used in Narendra and Parthasarathy [49] for testing the neural identifiers. The simulation results show that: (1) convergence of the back-propagation training algorithm for the fuzzy logic system was much faster than that of the back-propagation training algorithm for the neural network; (2) the fuzzy identifiers achieved similar or better performance than the neural identifiers, using simpler systems (in the sense of fewer adjustable parameters which need to be trained by the back-propagation algorithms); and (3) the fuzzy identifier converged to the true system faster after incorporating some linguistic information.

4

TRAINING OF FUZZY LOGIC SYSTEMS USING ORTHOGONAL LEAST SQUARES

4.1 INTRODUCTION

In Chapter 3 we represented the fuzzy logic systems as three-layer feedforward networks, and based on this representation, the back-propagation algorithm was developed to train the fuzzy logic system to match desired input-output pairs. Because the fuzzy logic system is nonlinear in its adjustable parameters, the back-propagation algorithm implements a nonlinear gradient optimization procedure and can be trapped at a local minimum and converges slowly [although much faster than a comparable back-propagation neural network (see Chapter 3)]. In this chapter, we fix some parameters of the fuzzy logic system such that the resulting fuzzy system is equivalent to a series expansion of some basis functions which are named *fuzzy basis functions*. This fuzzy basis function expansion is linear in its adjustable parameters; therefore, we can use the classical Gram-Schmidt orthogonal least squares (OLS) algorithm to determine the significant fuzzy basis functions and the remaining parameters. The OLS algorithm is a one-pass regression procedure and is therefore much faster than the back-propagation algorithm. Also, the OLS algorithm generates a robust fuzzy logic system which is not sensitive to noise in its inputs.

The most important advantage of using fuzzy basis functions rather than polynomials [Chen, Billings, and Luo, 6; Powell, 58], radial basis functions [Chen, Cowan, and Grant, 5; Powell, 59], neural networks [Rumelhart and McCleland, 61], and so on, is that a linguistic fuzzy IF-THEN rule is naturally related to a fuzzy basis function. Linguistic fuzzy IF-THEN rules can often be obtained from human experts who are familiar with the system under consideration. For example, pilots can describe properties of an aircraft by linguistic fuzzy IF-THEN rules, and experienced operators of power plants can provide operational instructions in the form of linguistic fuzzy IF-THEN rules, and so on. These linguistic rules are very important and often contain information which is not contained in the input-output pairs obtained by measuring the outputs of a system for test inputs because the test inputs may not be rich enough to excite all the modes of the system. Using fuzzy basis function expansions, we can easily combine two sets of fuzzy basis functions—one generated from input-output pairs using the OLS algorithm, and the other obtained from linguistic fuzzy IF-THEN rules—into a single fuzzy basis function expansion, which is therefore constructed using both numerical and linguistic information in a uniform fashion.

4.2 FUZZY SYSTEMS AS FUZZY BASIS FUNCTION EXPANSIONS

Definition 4.1. Consider the fuzzy logic systems in Lemma 2.3, that is, (2.46). Define *fuzzy basis functions* (FBF) as

$$p_j(\underline{x}) = \frac{\Pi_{i=1}^{n} \mu_{F_i^j}(x_i)}{\sum_{j=1}^{M} \Pi_{i=1}^{n} \mu_{F_i^j}(x_i)}, \quad j = 1, 2, \ldots, M \quad (4.1)$$

where $\mu_{F_i^j}(x_i) = a_i^j exp(-\frac{1}{2}(\frac{x_i - \bar{x}_i^j}{\sigma_i^j})^2)$ are Gaussian membership functions. Then the fuzzy logic system (2.46) is equivalent to an *FBF expansion*

$$f(\underline{x}) = \sum_{j=1}^{M} p_j(\underline{x}) \theta_j \quad (4.2)$$

where $\theta_j = \bar{y}^j \in R$ are constants.

From (4.1) and (2.16) we see that an FBF corresponds to a fuzzy IF-THEN rule. Specifically, an FBF for a rule can be determined as follows: first, calculate the product of all membership functions for the linguistic terms in the IF part of the rule, and call it a *pseudo-FBF* for the rule; then after calculating the pseudo-FBFs for all the M rules, the FBF for the jth rule is determined by dividing the pseudo-FBF for the jth rule by the sum of all the M pseudo-FBFs. An FBF can either be determined based on a given linguistic rule as previously, or generated based on a numerical input-output pair (as shown later in the Initial FBF Determination method described in Section 4.3).

What does the FBF of (4.1) look like when plotted as a function of $\underline{x}$? We now consider a simple one-dimensional example (that is, $n = 1$). Suppose that we have four fuzzy rules in the form of (2.16) with $\mu_{F^j}(x) = exp[-\frac{1}{2}(x - \bar{x}^j)^2]$, where $\bar{x}^j = -3, -1, 1, 3$ for $j = 1, 2, 3, 4$, respectively (note that the FBFs are determined only based on the IF parts of the rules, so we do not need the $\mu_{G^j}(z)$). Therefore, $p_j(x) = exp[-\frac{1}{2}(x - \bar{x}^j)^2]/\sum_{i=1}^{4} exp[-\frac{1}{2}(x - \bar{x}^i)^2]$, which are plotted in Figure 4.1 from left to right for $j = 1, 2, 3, 4$, respectively. From Figure 4.1 we see an interesting property of the FBFs: the $p_j(x)$'s whose centers $\bar{x}^j$ are

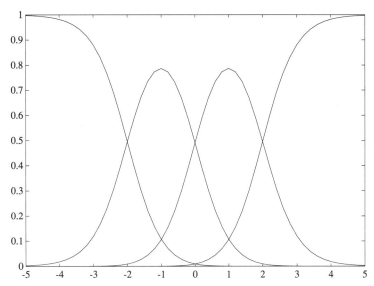

Figure 4.1 An example of the fuzzy basis functions.

inside the interval $[-3, 3]$ (which contains all the centers) look like Gaussian functions, whereas the $p_j(x)$'s whose centers $\bar{x}^j$ are on the boundaries of the interval $[-3, 3]$ look like sigmoidal functions [Cybenko, 12]. It is known in the neural network literature that Gaussian radial basis functions are good at characterizing local properties, whereas neural networks with sigmoidal nonlinearities are good at characterizing global properties [Lippmann, 36]. Our FBFs seem to combine the advantages of both the Gaussian radial basis functions and the sigmoidal neural networks. Specifically, for regions in the input space U which have sampling points (we often use the sampling points as centers of the FBFs; see Section 4.3), the FBFs cover them with Gaussian-like functions so that higher resolution can be obtained for the FBF expansion over these regions. On the other hand, for regions in U which have no sampling points, the FBFs cover them with sigmoidal-like functions which have been shown to have good global properties [Cybenko, 12; Lippmann, 36]. Of course, all the preceding instances are empirical observations; an interesting research topic would be to study the properties of the FBFs from a rigorous mathematical point of view.

Equation (4.1) defines only one kind of FBF, that is, it defines the FBF for fuzzy systems with center average defuzzifier, singleton fuzzifier, product inference, and Gaussian membership function. Other fuzzy systems can have other forms of FBFs; for example, the fuzzy systems with minimum inference have an FBF in the form of (4.1) with product operation replaced by minimum operation; however, the basic idea remains the same, that is, to view a fuzzy system as a linear combination of functions which are defined as FBFs. Different FBFs have different properties.

We can analyze (4.2) from two points of view. First, if we view all the parameters a_i^j, $\bar{x}_i^j$, and σ_i^j in $p_j(\underline{x})$ as free design parameters, then the FBF expansion (4.2) is nonlinear in the parameters. In order to specify such an FBF expansion, we must use nonlinear optimization techniques, for example, use the back-propagation algorithm of Chapter 3. On the other hand, we can fix all the parameters in $p_j(\underline{x})$ at the very beginning of the FBF expansion design procedure, so that the only free design parameters are θ_j; in this case, $f(\underline{x})$ of (4.2) is linear in the parameters. We adopt this second point of view in this chapter. The advantage of this point of view is that we are now able to use very efficient linear parameter estimation methods, for example, the Gram-Schmidt orthogonal least squares algorithm [Chen, Cowan, and Grant, 5; Chen, Billings, and Luo, 6] to design the FBF expansions.

4.3 ORTHOGONAL LEAST SQUARES LEARNING

In order to describe how the orthogonal least squares (OLS) learning algorithm works, we view the fuzzy basis function expansion (4.2) as a special case of the linear regression model

$$d(t) = \sum_{j=1}^{M} p_j(t)\theta_j + e(t) \quad (4.3)$$

where $d(t)$ is system output, θ_j are real parameters, $p_j(t)$ are known as regressors which are fixed functions of system inputs $\underline{x}(t)$, that is,

$$p_j(t) = p_j(\underline{x}(t)) \quad (4.4)$$

and $e(t)$ is an error signal which is assumed to be uncorrelated with the regressors. Suppose that we are given N input-output pairs: $(\underline{x}^0(t), d^0(t))$, $t = 1, 2, \ldots, N$. Our task is to design an FBF expansion $f(\underline{x})$ such that the error between $f(\underline{x}^0(t))$ and $d^0(t)$ is minimized.

In order to present the OLS algorithm, we arrange (4.3) from $t = 1$ to N in the following matrix form:

$$\underline{d} = P\underline{\theta} + \underline{e} \quad (4.5)$$

where $\underline{d} = [d(1), \ldots, d(N)]^T$, $P = [\underline{p}_1, \ldots, \underline{p}_M]$ with $\underline{p}_i = [p_i(1), \ldots, p_i(N)]^T$, $\underline{\theta} = [\theta_1, \ldots, \theta_M]^T$, and $\underline{e} = [e(1), \ldots, e(N)]^T$. The OLS algorithm transforms the set of $\underline{p}_i$ into a set of orthogonal basis vectors, and uses only the significant basis vectors to form the final FBF expansion. In order to perform the OLS procedure, we first need to fix the parameters a_i^j, $\bar{x}_i^j$, and σ_i^j in the FBF $p_j(\underline{x})$ based on the input-output pairs. We propose the following scheme:

Initial FBF Determination. Choose N initial $p_j(\underline{x})$'s in the form of (4.1) (for this case, M in (4.1) equals N), with the parameters determined as follows: $a_i^j = 1$, $\bar{x}_i^j = x_i^0(j)$, and $\sigma_i^j = [max(x_i^0(j), j = 1, 2, \ldots, N) - min(x_i^0(j), j = 1, 2, \ldots, N)]/M_s$, where $i = 1, 2, \ldots, n$, $j = 1, 2, \ldots, N$, and M_s is the number

Sec. 4.3 Orthogonal Least Squares Learning

of FBFs in the final FBF expansion. We assume that M_s is given based on practical constraints; in general, $M_s \ll N$.

We choose $a_i^j = 1$ because $\mu_{A_i^j}(x_i)$ are fuzzy membership functions which can be assumed to achieve unity membership value at some center $\bar{x}_i^j$. We choose the centers $\bar{x}_i^j$ to be the input points in the given input-output pairs. Finally, the earlier choice of σ_i^j should make the final FBFs "uniformly" cover the input region spanned by the input points in the given input-output pairs.

Next, we use the OLS algorithm, similar to that in Chen, Cowan, and Grant [5] and Chen, Billings, and Luo [6] (based on the classical Gram-Schmidt orthogonalization procedure) to select the significant FBFs from the N FBFs determined by the Initial FBF Determination method:

☐ At the first step, for $1 \leq i \leq N$, compute

$$\underline{w}_1^{(i)} = \underline{p}_i, \qquad g_1^{(i)} = \left(\underline{w}_1^{(i)}\right)^T \underline{d}^0 / \left(\left(\underline{w}_1^{(i)}\right)^T \underline{w}_1^{(i)}\right) \tag{4.6}$$

$$[err]_1^{(i)} = \left(g_1^{(i)}\right)^2 \left(\underline{w}_1^{(i)}\right)^T \underline{w}_1^{(i)} / (\underline{d}^{0T} \underline{d}^0) \tag{4.7}$$

where $\underline{p}_i = [p_i(\underline{x}^0(1)), \ldots, p_i(\underline{x}^0(N))]^T$, and $p_i(\underline{x}^0(t))$ are given by the Initial FBF Determination method. Find

$$[err]_1^{(i_1)} = max\left([err]_1^{(i)}, 1 \leq i \leq N\right) \tag{4.8}$$

and select

$$\underline{w}_1 = \underline{w}_1^{(i_1)} = \underline{p}_{i_1}, \qquad g_1 = g_1^{(i_1)} \tag{4.9}$$

☐ At the kth step where $2 \leq k \leq M_s$, for $1 \leq i \leq N, i \neq i_1, \ldots, i \neq i_{k-1}$, compute

$$\alpha_{jk}^{(i)} = \underline{w}_j^T \underline{p}_i / (\underline{w}_j^T \underline{w}_j), \qquad 1 \leq j < k \tag{4.10}$$

$$\underline{w}_k^{(i)} = \underline{p}_i - \sum_{j=1}^{k-1} \alpha_{jk}^{(i)} \underline{w}_j, \qquad g_k^{(i)} = (\underline{w}_k^{(i)})^T \underline{d}^0 / ((\underline{w}_k^{(i)})^T \underline{w}_k^{(i)}) \tag{4.11}$$

$$[err]_k^{(i)} = (g_k^{(i)})^2 (\underline{w}_k^{(i)})^T \underline{w}_k^{(i)} / (\underline{d}^{0T} \underline{d}^0) \tag{4.12}$$

Find

$$[err]_k^{(i_k)} = max([err]_k^{(i)}, 1 \leq i \leq N, i \neq i_1, \ldots, i \neq i_{k-1}) \tag{4.13}$$

and select

$$\underline{w}_k = \underline{w}_k^{(i_k)}, \qquad g_k = g_k^{(i_k)} \tag{4.14}$$

☐ Solve the triangular system

$$A^{(M_s)} \underline{\theta}^{(M_s)} = \underline{g}^{(M_s)} \tag{4.15}$$

where

$$A^{(M_s)} = \begin{bmatrix} 1 & \alpha_{12}^{(i_2)} & \alpha_{13}^{(i_3)} & \cdots & \alpha_{1M_s}^{(i_{M_s})} \\ 0 & 1 & \alpha_{23}^{(i_3)} & \cdots & \alpha_{2M_s}^{(i_{M_s})} \\ \cdots & \cdots & \cdots & \cdots & \cdots \\ 0 & 0 & \cdots & 1 & \alpha_{M_s-1,M_s}^{(i_{M_s})} \\ 0 & 0 & 0 & \cdots & 1 \end{bmatrix}, \quad (4.16)$$

$$\underline{g}^{(M_s)} = [g_1, \ldots, g_{M_s}]^T, \quad \underline{\theta}^{(M_s)} = [\theta_1^{(M_s)}, \ldots, \theta_{M_s}^{(M_s)}]^T \quad (4.17)$$

The final FBF expansion is

$$f(\underline{x}) = \sum_{j=1}^{M_s} p_{i_j}(\underline{x}) \theta_j^{(M_s)} \quad (4.18)$$

where $p_{i_j}(\underline{x})$ are the subset of the FBFs determined by the Initial FBF Determination method with i_j determined by the preceding steps.

Some comments on this OLS algorithm are now in order.

1. The purpose of the original Gram-Schmidt OLS algorithm is to perform an orthogonal decomposition for P, that is, $P = WA$, where W is an orthogonal matrix, and A is an upper-triangular matrix with unity diagonal elements. Substituting $P = WA$ into (4.5), we have that $\underline{d} = WA\underline{\theta} + \underline{e} = W\underline{g} + \underline{e}$, where $\underline{g} = A\underline{\theta}$ has the same meaning as used in our OLS algorithm, and the $\alpha_{jk}^{(i)}$ in our OLS algorithm corresponds to the elements of A. Our OLS algorithm does not complete the decomposition of $P = WA$, but only selects some domain columns from P.

2. The $[err]_k^{(i)} = (g_k^{(i)})^2 (\underline{w}_k^{(i)})^T \underline{w}_k^{(i)} / (\underline{d}^{0T} \underline{d}^0)$ represents the error-reduction ratio due to $\underline{w}_k^{(i)}$; hence, our OLS algorithm selects significant FBFs based on their error-reduction ratio. That is, the FBFs with largest error-reduction ratios are retained in the final FBF expansion.

4.4 CONTROL OF THE NONLINEAR BALL-AND-BEAM SYSTEM USING FBF EXPANSIONS

In this section, we use the OLS algorithm to design an FBF expansion to approximate a controller for the nonlinear ball-and-beam system Hauser, Sastry, and Kokotovic [18]. Our purpose is to use the FBF expansion as a controller to regulate the system to the origin from a certain range of initial conditions. We first use the input-output linearization algorithm of Hauser, Sastry, and Kokotovic [18] to generate a set of state-control pairs for randomly sampled points in a certain region of the state space, and then view these state-control pairs as the input-output pairs in Section 4.3 and use the OLS algorithm to determine an FBF expansion which is used as the controller for the ball-and-beam system with initial conditions arbitrarily chosen in the sampled state space. In other words, we use the controller of Hauser, Sastry, and Kokotovic [18] to generate a look-up table of state-control pairs, and then use the FBF expansion to interpolate these pairs to form the final controller. For many practical problems, this kind of look-up table of state-control pairs can be provided by human experts or collected from past successful control executions.

The ball-and-beam system, which can be found in many undergraduate control laboratories, is shown in Figure 4.2. The beam is made to rotate in a vertical plane by applying a torque at the center of rotation and the ball is free to roll along the beam. We require that the ball remain in contact with the beam. Let $\underline{x} = (r, \dot{r}, \theta, \dot{\theta})^T$ be the state vector of the system and $y = r$ be the output of the system. Then, from Hauser, Sastry, and Kokotovic [18], the system can be represented by the state-space model

$$\begin{bmatrix} \dot{x}_1 \\ \dot{x}_2 \\ \dot{x}_3 \\ \dot{x}_4 \end{bmatrix} = \begin{bmatrix} x_2 \\ B(x_1 x_4^2 - G\sin x_3) \\ x_4 \\ 0 \end{bmatrix} + \begin{bmatrix} 0 \\ 0 \\ 0 \\ 1 \end{bmatrix} u \qquad (4.19)$$

$$y = x_1 \qquad (4.20)$$

where the control u is the acceleration of θ, and the parameters B and G are defined in Hauser, Sastry, and Kokotovic [18]. The purpose of control is to determine $u(\underline{x})$ such that the closed-loop system output y will converge to zero from arbitrary initial conditions in a certain region.

The input-output linearization algorithm of Hauser, Sastry, and Kokotovic [18] determines the control law $u(\underline{x})$ as follows: for state $\underline{x}$, compute $v(\underline{x}) = -\alpha_3 \phi_4(\underline{x}) - \alpha_2 \phi_3(\underline{x}) - \alpha_1 \phi_2(\underline{x}) - \alpha_0 \phi_1(\underline{x})$, where $\phi_1(\underline{x}) = x_1$, $\phi_2(\underline{x}) = x_2$, $\phi_3(\underline{x}) = -BG\sin x_3$, $\phi_4(\underline{x}) = -BGx_4\cos x_3$, and the α_i are chosen so that $s^4 + \alpha_3 s^3 + \alpha_2 s^2 + \alpha_1 s + \alpha_0$ is a Hurwitz polynomial; compute $a(\underline{x}) = -BG\cos x_3$ and $b(\underline{x}) = BGx_4^2 \sin x_3$; then, $u(\underline{x}) = (v(\underline{x}) - b(\underline{x}))/a(\underline{x})$.

In our simulations, we used the $u(\underline{x}) = (v(\underline{x}) - b(\underline{x}))/a(\underline{x})$ to generate N $(\underline{x}, u)$ pairs with $\underline{x}$ randomly sampled in the region $U = [-5, 5] \times [-2, 2] \times [-\pi/4, \pi/4] \times [-0.8, 0.8]$. We simulated three cases: Case 1: $N = 200$, $M_s = 20$, and the final FBF expansion $f(\underline{x})$ of (4.18) was used as the control u in (4.19); Case 2: $N = 40$, $M_s = 20$, and the final FBF expansion $f(\underline{x})$ of (4.18) was used

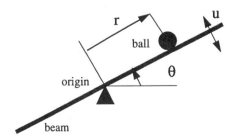

Figure 4.2 The ball-and-beam system.

as the control u in (4.19); and Case 3: $N = 40$, $M_s = 20$, and the control

$$u(\underline{x}) = \frac{1}{2}[f(\underline{x}) + f^L(\underline{x})] \tag{4.21}$$

where $f(\underline{x})$ is given by (4.18), and $f^L(\underline{x})$ is a *linguistic controller* which is in the form of (2.40) as determined based on the following four common sense linguistic control rules:

R_1^L : IF x_1 is positive and x_2 is near zero and x_3 is positive

and x_4 is near zero, THEN u is negative (4.22)

R_2^L : IF x_1 is positive and x_2 is near zero and x_3 is negative

and x_4 is near zero, THEN u is positive big (4.23)

R_3^L : IF x_1 is negative and x_2 is near zero and x_3 is positive

and x_4 is near zero, THEN u is negative big (4.24)

R_4^L : IF x_1 is negative and x_2 is near zero and x_3 is negative

and x_4 is near zero, THEN u is positive (4.25)

where the *positive* for x_1 is a fuzzy set $P1$ with membership function $\mu_{P1}(x_1) = exp[-\frac{1}{2}(\frac{min(x_1 - 4.0)}{4})^2]$; the *negative* for x_1 is a fuzzy set $N1$ with membership function $\mu_{N1}(x_1) = exp[-\frac{1}{2}(\frac{max(x_1 + 4.0)}{4})^2]$; the *near zero* for both x_2 and x_4 is a fuzzy set ZO with $\mu_{ZO}(x) = exp[-\frac{1}{2}x^2]$, the *positive* for x_3 is a fuzzy set $P3$ with $\mu_{P3}(x_3) = exp[-\frac{1}{2}(\frac{min(x_3 - \pi/4.0)}{\pi/4})^2]$; the *negative* for x_3 is a fuzzy set $N3$ with $\mu_{N3}(x_3) = exp[-\frac{1}{2}(\frac{max(x_3 + \pi/4.0)}{\pi/4})^2]$; the *positive* for u is a fuzzy set Pu with $\mu_{Pu}(u) = exp[-\frac{1}{2}(u - 0.1)^2]$; the *negative* for u is a fuzzy set Nu with $\mu_{Nu}(u) = exp[-\frac{1}{2}(u + 0.1)^2]$; the *positive big* for u is a fuzzy set PBu with $\mu_{PBu}(u) = exp[-\frac{1}{2}(u - 0.4)^2]$; and the *negative big* for u is a fuzzy set NBu with $\mu_{NBu}(u) = exp[-\frac{1}{2}(u + 0.4)^2]$. The preceding membership functions for the IF parts of R_1^L–R_4^L were determined based on the meaning of the linguistic terms; the parameters of the THEN part membership functions were determined by common sense and trial and error. The detailed formula of $f^L(\underline{x})$ can be easily obtained based on (2.40) and the preceding membership functions.

Clearly, R_1^L–R_4^L are determined based on our common sense of how to control the ball to stay at the origin when the ball is in certain regions. Take R_1^L as an example. If the ball stays at its position depicted in Figure 4.2 (which just corresponds to the IF part of R_1^L), then we should move the beam downward to reduce θ (but not a lot), which is equivalent to saying "u is negative" because the control u equals the acceleration of θ (see (4.19)). Although these common sense control rules are not precise, the control performance will, as we show later, be greatly improved by incorporating them into the controller (4.21).

We simulated each of the three cases for four initial conditions, $\underline{x}(0) = [2.4, -0.1, 0.6, 0.1]^T$, $[1.6, 0.05, -0.6, -0.05]^T$, $[-1.6, -0.05, 0.6, 0.05]^T$, and

$[-2.4, 0.1, -0.6, -0.1]^T$, which were arbitrarily chosen in $U = [-5, 5] \times [-2, 2] \times [-\pi/4, \pi/4] \times [-0.8, 0.8]$. Figures 4.3–4.5 depict the output y of the closed-loop system for Cases 1–3, respectively. In the simulations, we solved the differential equations using the MATLAB command "ode23" which uses the second-order and third-order Runge-Kutta method.

Some comments on these simulation results are now in order. (1) The fuzzy controller in Case 1 gave the best overall performance; this suggests that given a sufficient number of state-control pairs, the OLS algorithm can determine a successful FBF expansion controller. (2) The fuzzy controller in Case 2 could regulate the ball to the origin for some initial conditions, but the closed-loop system was unstable for some initial condition. This suggests that a sufficient sampling of the state space is important for the "pure numerical" fuzzy controller to be successful. (3) By using the same small number of state-control pairs but adding the fuzzy control rules (4.22)–(4.25), the fuzzy controller in Case 3 showed much better performance than the fuzzy controller in Case 2; that is, control performance was greatly improved by incorporating (in the sense of (4.21)) the linguistic fuzzy control rules into the controller.

We also simulated two extreme cases: (1) using the original controller of Hauser, Sastry, and Kokotovic [18], and (2) using only the pure linguistic controller $f^L(\underline{x})$ based on $R_1^L - R_4^L$, for the same initial conditions as in Cases 1–3. Figures 4.6 and 4.7 show the output of the closed-loop system for the first and second cases, respectively. Comparing Figure 4.6 with Figures 4.3–4.5 we see that

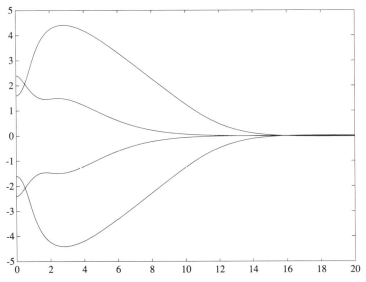

Figure 4.3 Outputs $r(t)$ of the closed-loop ball-and-beam system for Case 1 and four initial conditions.

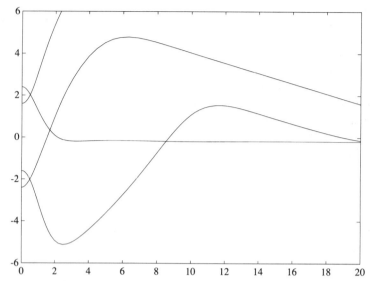

Figure 4.4 Outputs $r(t)$ of the closed-loop ball-and-beam system for Case 2 and four initial conditions.

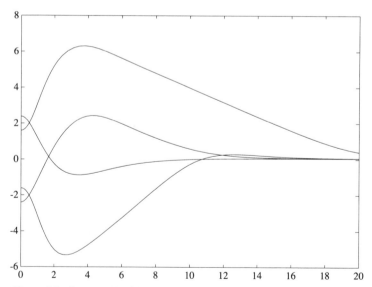

Figure 4.5 Outputs $r(t)$ of the closed-loop ball-and-beam system for Case 3 and four initial conditions.

Sec. 4.4 Control of the Nonlinear Ball-and-Beam System Using FBF Expansions 59

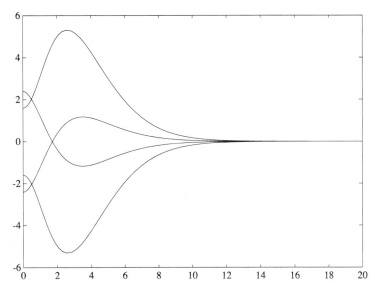

Figure 4.6 Outputs $r(t)$ of the closed-loop ball-and-beam system using the input-output linearization algorithm of Hauser, Sastry, and Kokotovic [18] and four initial conditions.

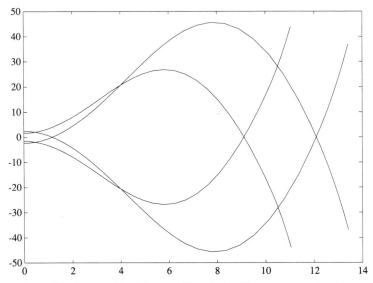

Figure 4.7 Outputs $r(t)$ of the closed-loop ball-and-beam system using the pure fuzzy logic controller based on the four linguistic rules (4.22)–(4.25) and four initial conditions.

the original controller of Hauser, Sastry, and Kokotovic [18] gave the best performance; this is to be expected because we used the FBF expansions to approximate this controller. Figure 4.7 shows that if we use the FBF expansion controller based only on the four linguistic rules (4.22)–(4.25), the closed-loop system is unstable; that is, the pure fuzzy logic controller with only four linguistic rules is not sufficient to control the system.

4.5 MODELING THE MACKEY-GLASS CHAOTIC TIME SERIES BY FBF EXPANSION

Time-series prediction is an important practical problem [Box and Jenkins, 4]. Applications of time-series prediction can be found in the areas of economic and business planning, inventory and production control, weather forecasting, signal processing, control, and many other fields. Here we apply our OLS approach to the prediction of the Mackey-Glass chaotic time series. Chaotic time series are generated from deterministic nonlinear systems and are sufficiently complicated that they appear to be "random" time series; however, because there are underlying deterministic maps that generate the series, chaotic time series are not random time series [Holden, 19].

The Mackey-Glass chaotic time series is generated from the following delay differential equation:

$$\frac{dx(t)}{dt} = \frac{0.2x(t-\tau)}{1+x^{10}(t-\tau)} - 0.1x(t) \quad (4.26)$$

When $\tau > 17$, Eq. (4.26) shows chaotic behavior. Higher values of τ yield higher dimensional chaos. In our simulation, we chose the series with $\tau = 30$. Figure 4.8 shows a section of 1,000 points (from $t = 1,001$ to 2,000) of the Mackey-Glass chaotic time series. Our task is to determine an FBF expansion as a one-step-ahead predictor for the chaotic time series based on the data section of Figure 4.8; that is, we will determine

$$\hat{x}(t) = f(\underline{z}(t)) \quad (4.27)$$

where $f(*)$ is in the form of (4.18), and $\underline{z}(t) = [x(t-1), \ldots, x(t-n)]^T$ are past observations of the series.

We chose $M_s = 80$ and $n = 10$. The residuals determined by

$$e(t) = x(t) - \hat{x}(t) \quad (4.28)$$

are shown in Figure 4.9. The series $x(t)$ and $\hat{x}(t)$ look almost indistinguishable if we depict them on the same plot.

We performed two kinds of statistical tests—autocorrelation test and chi-squared test—in order to determine whether the final FBF expansion is a valid model for the chaotic time series. The autocorrelations of $e(t)$ are plotted in

Sec. 4.5   Modeling the Mackey-Glass Chaotic Time Series by FBF Expansion   61

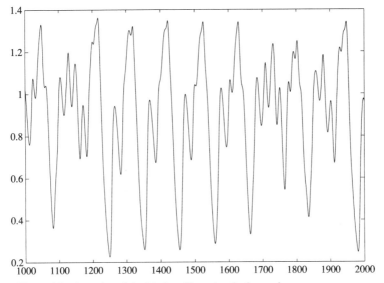

Figure 4.8 A section of the Mackey-Glass chaotic time series.

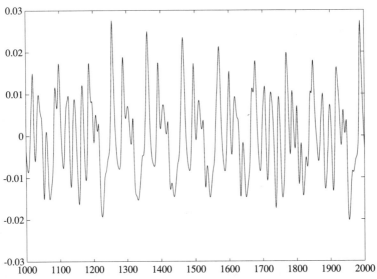

Figure 4.9 Residual sequence using the FBF expansion predictor for the chaotic time series of Figure 4.8.

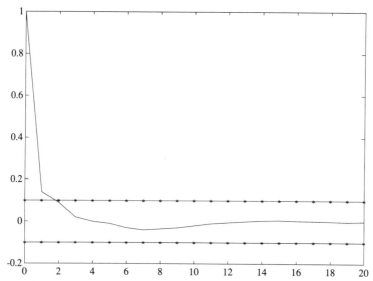

Figure 4.10 Autocorrelations of the residuals of Figure 4.9. -*- represents 95 percent confidence band.

Figure 4.10. To perform the chi-squared test, define

$$\underline{r}(t) = [w(t), \ldots, w(t - \eta + 1)]^T \qquad (4.29)$$

where $w(t)$ is some function of the past observations and residuals, and let

$$\Gamma^T \Gamma = N^{-1} \sum_{t=1,001}^{N+1,000} \underline{r}(t)\underline{r}^T(t) \qquad (4.30)$$

where $N = 1,000$. The chi-squared statistics are calculated according to Chen, Cowan, and Grant [5] as

$$\zeta(\eta) = N\underline{\mu}^T (\Gamma^T \Gamma)^{-1} \underline{\mu} \qquad (4.31)$$

where

$$\underline{\mu} = N^{-1} \sum_{t=1,001}^{N+1,000} \underline{r}(t)e(t)/\sigma_e \qquad (4.32)$$

and σ_e^2 is the variance of the residuals $e(t)$. Figures 4.11 and 4.12 show the chi-squared statistics $\zeta(\eta)$ for $w(t) = e^2(t-1)$ and $w(t) = e^2(t-1)y^2(t-1)$, respectively; we see that they are all within the 95 percent confidence band. Consequently, the model validity tests confirm that this FBF expansion is an adequate model for the chaotic time series.

Sec. 4.5   Modeling the Mackey-Glass Chaotic Time Series by FBF Expansion   63

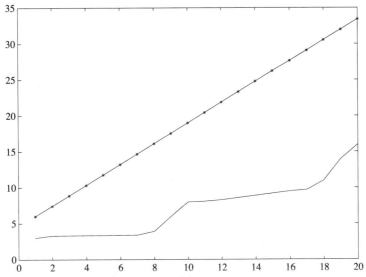

Figure 4.11   Chi-squared statistics $\zeta(\eta)$ for $w(t) = e^2(t-1)$. -*- represents 95 percent confidence band.

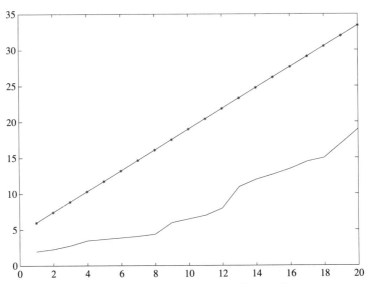

Figure 4.12   Chi-squared statistics $\zeta(\eta)$ for $w(t) = e^2(t-1)y^2(t-1)$. -*- represents 95 percent confidence band.

4.6 CONCLUDING REMARKS

In this chapter, we showed that fuzzy systems can be represented as linear combinations of fuzzy basis functions, developed an orthogonal least squares algorithm to select the significant fuzzy basis functions, and used the fuzzy basis function expansions as controllers for the ball-and-beam system and as predictors for the Mackey-Glass chaotic time series. Through a simple example we illustrated that the fuzzy basis functions whose centers are inside the sampling region look Gaussian, whereas the fuzzy basis functions whose centers are on the boundaries of the sampling regions look sigmoidal. The most important advantage of the fuzzy basis functions is that a linguistic fuzzy IF-THEN rule is directly related to a fuzzy basis function, so that the fuzzy basis function expansion provides a natural framework to combine both numerical information (in the form of input-output pairs) and linguistic information (in the form of fuzzy IF-THEN rules) in a uniform fashion. We showed an example of how to combine the fuzzy basis functions generated from a numerical state-control table and the fuzzy basis functions generated from some common sense linguistic fuzzy control rules to form a controller for the nonlinear ball-and-beam system. The simulation results showed that the control performance was greatly improved by incorporating these linguistic fuzzy control rules.

5

TRAINING OF FUZZY LOGIC SYSTEMS USING A TABLE-LOOKUP SCHEME

5.1 INTRODUCTION

Although the two training algorithms in Chapters 3 and 4 have well-justified criteria and work well for a number of examples, they are not simple in the sense that they may require intensive computations, since the back-propagation algorithm performs a nonlinear search procedure and the orthogonal least squares algorithm needs iterative operations. In this chapter, we develop a very simple method for adaptive fuzzy system design which performs a one-pass operation on the numerical input-output pairs and linguistic fuzzy IF-THEN rules. The key idea of this method is to generate fuzzy rules from input-output pairs, collect the generated rules and linguistic rules into a common fuzzy rule base, and construct a final fuzzy logic system based on the combined fuzzy rule base.

5.2 GENERATING FUZZY RULES FROM NUMERICAL DATA

Suppose we are given a set of desired input-output data pairs:

$$(x_1^{(1)}, x_2^{(1)}; y^{(1)}), (x_1^{(2)}, x_2^{(2)}; y^{(2)}), \cdots \qquad (5.1)$$

where x_1 and x_2 are inputs, and y is the output. This simple two-input one-output case is chosen in order to emphasize and to clarify the basic ideas of our new approach; extensions to general multi-input–multi-output cases are straightforward and will be discussed later in this section. The task here is to generate a set of fuzzy IF-THEN rules from the desired input-output pairs of (5.1), and use these fuzzy IF-THEN rules to determine a fuzzy logic system $f : (x_1, x_2) \to y$.

This approach consists of the following five steps:

Step 1: Divide the Input and Output Spaces into Fuzzy Regions

Assume that the domain intervals of x_1, x_2, and y are $[x_1^-, x_1^+]$, $[x_2^-, x_2^+]$, and $[y^-, y^+]$, respectively, where *domain interval* of a variable means that most probably this variable will lie in this interval (the values of a variable are allowed to lie outside its domain interval). Divide each domain interval into $2N+1$ regions (N can be different for different variables, and the lengths of these regions can be equal or unequal), denoted by SN (Small N), ..., $S1$ (Small 1), CE (Center), $B1$ (Big 1), ..., BN (Big N), and assign each region a fuzzy membership function. Figure 5.1 shows an example where the domain interval of x_1 is divided into five regions ($N = 2$), the domain region of x_2 is divided into seven regions ($N = 3$), and the domain interval of y is divided into five regions ($N = 2$). The shape of each membership function is triangular; one vertex lies at the center of the region and has membership value unity; the other two vertices lie at the centers of the two neighboring regions, respectively, and have membership values equal to zero. Of course, other divisions of the domain regions and other shapes of membership functions are possible.

Step 2: Generate Fuzzy Rules from Given Data Pairs

First, *determine the degrees of given $x_1^{(i)}$, $x_2^{(i)}$, and $y^{(i)}$ in different regions*. For example, $x_1^{(1)}$ in Figure 5.1 has degree 0.8 in $B1$, degree 0.2 in $B2$, and zero degrees in all other regions. Similarly, $x_2^{(2)}$ in Figure 5.1 has degree 1 in CE, and zero degrees in all other regions.

Second, *assign a given $x_1^{(i)}$, $x_2^{(i)}$, or $y^{(i)}$ to the region with maximum degree*. For example, $x_1^{(1)}$ in Figure 5.1 is considered to be $B1$, and $x_2^{(2)}$ in Figure 5.1 is considered to be CE.

Finally, *obtain one rule from one pair of desired input-output data*, for example,
$(x_1^{(1)}, x_2^{(1)}; y^{(1)}) \Rightarrow [x_1^{(1)}$ (0.8 in $B1$, max), $x_2^{(1)}$ (0.7 in $S1$, max); $y^{(1)}$ (0.9 in CE, max)] $\Rightarrow$ Rule 1: IF x_1 is $B1$ and x_2 is $S1$, THEN y is CE;
$(x_1^{(2)}, x_2^{(2)}; y^{(2)}) \Rightarrow [x_1^{(2)}$ (0.6 in $B1$, max), $x_2^{(2)}$ (1 in CE, max); $y^{(2)}$ (0.7 in $B1$, max)] $\Rightarrow$ Rule 2: IF x_1 is $B1$ and x_2 is CE, THEN y is $B1$.

The rules generated in this way are "and" rules, that is, rules in which the conditions of the IF part must be met simultaneously in order for the result of the THEN part to occur.

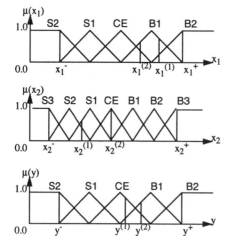

Figure 5.1 Divisions of the input and output spaces into fuzzy regions and the corresponding membership functions.

Sec. 5.2 Generating Fuzzy Rules from Numerical Data

For the problems considered here, that is, generating fuzzy rules from numerical data, only "and" rules are required since the antecedents are different components of a single input vector.

Step 3: Assign a Degree to Each Rule

Since there are usually many data pairs and each data pair generates one rule, it is highly probable that there will be some conflicting rules, that is, rules which have the same IF part but a different THEN part. One way to resolve this conflict is to assign a degree to each rule generated from data pairs and accept only the rule from a conflict group that has maximum degree. In this way not only is the conflict problem resolved, but also the number of rules is greatly reduced.

We use the following product strategy to assign a degree to each rule: for the rule "IF x_1 is A and x_2 is B, THEN y is C," the *degree of this rule*, denoted by $D(Rule)$, is defined as

$$D(Rule) = \mu_A(x_1)\mu_B(x_2)\mu_C(y) \tag{5.2}$$

As examples, Rule 1 has degree

$$\begin{aligned} D(Rule1) &= \mu_{B1}(x_1)\mu_{S1}(x_2)\mu_{CE}(y) \\ &= 0.8 \times 0.7 \times 0.9 = 0.504 \end{aligned} \tag{5.3}$$

(see Figure 5.1) and Rule 2 has degree

$$\begin{aligned} D(Rule2) &= \mu_{B1}(x_1)\mu_{CE}(x_2)\mu_{B1}(y) \\ &= 0.6 \times 1 \times 0.7 = 0.42 \end{aligned} \tag{5.4}$$

In practice, we often have a priori information about the data pairs. For example, if we let an expert check given data pairs, the expert may suggest that some are very useful and crucial, but others are very unlikely and may be caused just by measurement errors. We can therefore assign a degree to each data pair which represents our belief of its usefulness.

Suppose the data pair $(x_1^{(1)}, x_2^{(1)}; y^{(1)})$ has degree $\mu^{(1)}$, then we redefine the degree of Rule 1 as

$$D(Rule1) = \mu_{B1}(x_1)\mu_{S1}(x_2)\mu_{CE}(y)\mu^{(1)} \tag{5.5}$$

That is, the degree of a rule is defined as the product of the degrees of its components and the degree of the data pair which generates this rule. This is important in practical applications because real numerical data have different reliabilities. For good data we assign higher degrees, and for bad data we assign lower degrees. In this way, human experience about the data is used in a common base as other information. If one emphasizes objectivity and does not want a human to judge the numerical data, our strategy still works by setting all the degrees of the data pairs equal to unity.

Step 4: Create a Combined Fuzzy Rule Base

Figure 5.2 illustrates a table-lookup representation of a fuzzy rule base. We fill the boxes of the base with fuzzy rules according to the following strategy: *a combined fuzzy rule base is assigned rules from either those generated from numerical data or linguistic rules (we assume that a linguistic rule also has a degree which is assigned by the human expert and reflects the expert's belief of the importance of the rule); if there is more than one rule in one box of the fuzzy rule base, use the rule that has maximum degree.* In this way, both

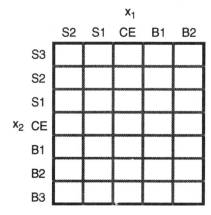

Figure 5.2 Table-lookup illustration of a fuzzy rule base.

numerical and linguistic information are codified into a common framework—the combined fuzzy rule base. If a linguistic rule is an "and" rule, it fills only one box of the fuzzy rule base, but if a linguistic rule is an "or" rule (that is, a rule for which the THEN part follows if any condition of the IF part is satisfied), it fills all the boxes in the rows or columns corresponding to the regions of the IF part. For example, suppose we have the linguistic rule: "IF x_1 is $S1$ or x_2 is CE, THEN y is $B2$" for the fuzzy rule base of Figure 5.2; then we fill the seven boxes in the column of $S1$ and the five boxes in the row of CE with $B2$. The degrees of all the $B2$'s in these boxes equal the degree of this "or" rule.

Step 5: Determine a Mapping Based on the Combined Fuzzy Rule Base

We use the following defuzzification strategy to determine the output control y for given inputs (x_1, x_2). First, for given inputs (x_1, x_2), we combine the antecedents of the ith fuzzy rule using *product* operations to determine the degree, $\mu^i_{O^i}$, of the output control corresponding to (x_1, x_2); that is,

$$\mu^i_{O^i} = \mu_{I^i_1}(x_1)\mu_{I^i_2}(x_2) \tag{5.6}$$

where O^i denotes the output region of Rule i, and I^i_j denotes the input region of Rule i for the jth component; for example, Rule 1 gives

$$\mu^1_{CE} = \mu_{B1}(x_1)\mu_{S1}(x_2) \tag{5.7}$$

Then we use the center average defuzzification formula to determine the output

$$y = \frac{\sum_{i=1}^{M} \mu^i_{O^i} \bar{y}^i}{\sum_{i=1}^{M} \mu^i_{O^i}} \tag{5.8}$$

where $\bar{y}^i$ denotes the center value of region O^i (the *center* of a fuzzy region is defined as the point which has the smallest absolute value among all the points at which the membership function for this region has membership value equal to 1), and M is the number of fuzzy rules in the combined fuzzy rule base.

From Steps 1 to 5 we see that our new method is simple and straightforward in the sense that it is a one-pass build-up procedure that does not require time-consuming training; hence, it has the same advantage that the fuzzy approach has over the neural approach, namely, it is simple and quick to construct.

This five-step procedure can easily be extended to general multi-input–multi-output cases. Steps 1 to 4 are independent of how many inputs and how many outputs there are. In Step 5, we only need to replace $\mu^i_{O^i}$ in Eq. (5.6) with $\mu^i_{O^i_j}$, where j denotes the jth component of the output vector (O^i_j is the region of Rule i for the jth output component; $\mu^i_{O^i_j}$ is the same for all j), and change Eq. (5.8) to

$$y_j = \frac{\sum_{i=1}^{\mu} \mu^i_{O^i_j} \bar{y}^i_j}{\sum_{i=1}^{\mu} \mu^i_{O^i_j}} \qquad (5.9)$$

where $\bar{y}^i_j$ denotes the center of region O^i_j.

5.3 APPLICATION TO TRUCK BACKER-UPPER CONTROL

Backing up a truck to a loading dock is a nonlinear control problem. A neural network controller for the truck backer-upper problem was developed in Nguyen and Widrow [51], whereas a fuzzy control strategy for the same problem was proposed in Kong and Kosko [28]. The neural network controller of Nguyen and Widrow [51] only uses numerical data and cannot utilize linguistic rules determined from expert drivers; on the other hand, the fuzzy controller of Kong and Kosko [28] only uses linguistic rules and cannot utilize sampled data. Since the truck backer-upper control problem is a good example of the kind of control system design problems that replace a human controller with a machine, it is interesting to apply the approach developed in Section 5.2 to this problem. In order to distinguish these methods, we call the method of Kong and Kosko [28] the *fuzzy approach*, the method of Nguyen and Widrow [51] the *neural approach*, and our method in this chapter the *numerical-fuzzy approach*.

The results of Kong and Kosko [28] demonstrated superior performance of the fuzzy controller over the neural controller; however, the fuzzy and neural controllers use different information to construct the control strategies. It is possible that the fuzzy rules used in Kong and Kosko [28] to construct the controller are more complete and contain more information than the numerical data used to construct the neural controller; hence, the comparison between the fuzzy and neural controllers, from a final control performance point of view, is somewhat unfair. If the linguistic fuzzy rules were incomplete and the numerical information contained many very good data pairs, it is highly possible that the neural controller would outperform the fuzzy controller.

Our numerical-fuzzy approach provides a fair basis for comparing fuzzy and neural controllers (the numerical-fuzzy approach can be viewed as a fuzzy approach in the sense that it differs from the pure fuzzy approach only in the way it obtains fuzzy IF-THEN rules). We can provide the same desired input-output pairs to both the neural and numerical-fuzzy approaches; consequently, we can compare the final control performances of both controllers fairly since they both use the same information.

Example 5.1. In this example, we use the same set of desired input-output pairs to simulate neural and numerical-fuzzy controllers, and compare their final control performances.

Statement of the Truck Backer-Upper Control Problem

The simulated truck and loading zone are shown in Figure 5.3. The truck corresponds to the cab part of the neural truck in the Nguyen-Widrow [51] neural truck backer-upper system. The truck position is exactly determined by the three state variables ϕ, x, and y, where ϕ is the angle of the truck with the horizontal as shown in Figure 5.3. Control to the truck is the angle θ. Only backing up is considered. The truck moves backward by a fixed unit distance every stage. For simplicity, we assume enough clearance between the truck and the loading dock such that y does not have to be considered as an input. The task here is to design a control system, whose inputs are $\phi \in [-90°, 270°]$ and $x \in [0, 20]$, and whose output is $\theta \in [-40°, 40°]$, such that the final states will be $(x_f, \phi_f) = (10, 90°)$.

Generating Desired Input-Output Pairs $(x, \phi; \theta)$

We do this by trial and error: at every stage (given ϕ and x) starting from an initial state, we determined a control θ based on common sense (that is, our own experience of how to control the steering angle in the situation); after some trials, we chose the desired input-output pairs corresponding to the smoothest successful trajectory.

The following 14 initial states were used to generate desired input-output pairs: (x_0, ϕ_0^o) = (1,0), (1,90), (1,270); (7,0), (7,90), (7,180), (7,270); (13,0), (13,90), (13,180), (13,270); (19,90), (19,180), (19,270). Since we performed simulations, we needed to know the dynamics of the truck backer-upper procedure. We used the following approximate kinematics (see Wang and Mendel [83]

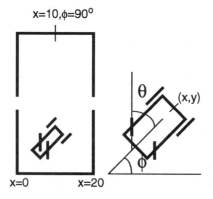

Figure 5.3 Diagram of simulated truck and loading zone.

for details):

$$x(t+1) = x(t) + cos[\phi(t) + \theta(t)] + sin[\theta(t)]sin[\phi(t)] \quad (5.10)$$

$$y(t+1) = y(t) + sin[\phi(t) + \theta(t)] - sin[\theta(t)]cos[\phi(t)] \quad (5.11)$$

$$\phi(t+1) = \phi(t) - sin^{-1}\left[\frac{2sin(\theta(t))}{b}\right] \quad (5.12)$$

where b is the length of the truck. We assumed $b = 4$ in the simulations of this section. Equations (5.10) to (5.12) were used to obtain the next state when the present state and control are given. Since y is not considered a state, only Eqs. (5.10) and (5.12) were used in the simulations. We wrote Eq. (5.11) here for the purpose of showing the complete dynamics of the truck. Observe, from Eqs. (5.10)–(5.12), that even this simplified dynamic model of the truck is nonlinear. The 14 sequences of desired $(x, \phi; \theta)$ pairs are given in Wang and Mendel [83]; we only include one such sequence in Table 5.1.

Neural Control and Simulation Results

We used a two-input single-output three-layer back-propagation neural network [Rumelhart and McCleland, 61] for our control task. Twenty hidden neurons were used, and a sigmoid non-linear function was used for each neuron. The

TABLE 5.1 Desired trajectory starting from $(x_0, \phi_0) = (1, 0°)$

t	x	$\phi°$	$\theta°$
0	1.00	0.00	−19.00
1	1.95	9.37	−17.95
2	2.88	18.23	−16.90
3	3.79	26.59	−15.85
4	4.65	34.44	−14.80
5	5.45	41.78	−13.75
6	6.18	48.60	−12.70
7	7.48	54.91	−11.65
8	7.99	60.71	−10.60
9	8.72	65.99	−9.55
10	9.01	70.75	−8.50
11	9.28	74.98	−7.45
12	9.46	78.70	−6.40
13	9.59	81.90	−5.34
14	9.72	84.57	−4.30
15	9.81	86.72	−3.25
16	9.88	88.34	−2.20
17	9.91	89.44	0.00
18			
19			
20			

output of the third-layer neuron represents the steering angle θ according to a uniform mapping from [0,1] to $[-40^o, 40^o]$. That is, if the neuron output is $g(t)$, the corresponding output $\theta(t)$ is

$$\theta(t) = 80g(t) - 40 \tag{5.13}$$

In the simulations, we normalized $[-40^o, 40^o]$ into $[-1, 1]$. Similarly, the inputs to the neurons were also normalized into $[-1, 1]$.

Our neural network controller is different from the Nguyen-Widrow neural controller of Nguyen and Widrow [51]. First, we have only one neural network which does the same work as the truck controller of the Nguyen-Widrow network; the truck emulator of the Nguyen-Widrow network is not needed in our task. Second, and more fundamentally, we train our neural network using desired input-output (state-control) pairs, which are obtained from the past successful control history of the truck, whereas Nguyen and Widrow [51] connect their neural network stage by stage and train these concatenated neural networks by back-propagating the error at the final state through this long network chain (the detailed algorithm is different from the standard error back-propagation algorithm in order to meet the constraint that the neural networks at each stage perform the same transformation; for details see Nguyen and Widrow, [51]). Hence, the training of our neural network is simpler than that of the Nguyen-Widrow network. Of course, we need to know some successful control trajectories (state-control pairs) starting from typical initial states; this is not required in the Nguyen-Widrow neural network controller.

We trained the neural network using the standard error back-propagation algorithm (see Rumelhart and McCleland [61] or Chapter 7) for the generated 14 sequences of desired $(x, \phi; \theta)$ pairs. We used the converged network to control the truck whose dynamics are approximately given by Eqs. (5.10)–(5.12). Three arbitrarily chosen initial states, (x_0, ϕ_0^o) = (3,-30), (10,220), and (13,30), were used to test the neural controller. The truck trajectories from the three initial states are shown in Figure 5.4. We see that the neural controller successfully controls the truck to the desired position starting from all three initial states.

Numerical-Fuzzy Control and Simulation Results

We used the five-step procedure of Section 5.2 to determine the control law $f : (x, \phi) \to \theta$, based on the 14 generated sequences of successful $(x, \phi; \theta)$ pairs. For this specific problem, we used membership functions shown in Figure 5.5. The fuzzy rules generated from the desired input-output pairs and their corresponding degrees are given in Wang and Mendel [83]; we only show the generated rules for the data pairs of Table 5.1 in Table 5.2. The final fuzzy rule base is shown in Figure 5.6. (This is the result of Step 4 of our method in Section 5.2; here we assume that no linguistic rules are available.) We see from Figure 5.6 that there are no generated rules for some ranges of x and ϕ. This shows that the desired trajectories from the 14 initial states do not cover all the possible cases; however, we will see that the rules in Figure 5.6 are sufficient for controlling the truck to the desired state starting from given initial states.

Sec. 5.3   Application to Truck Backer-Upper Control   73

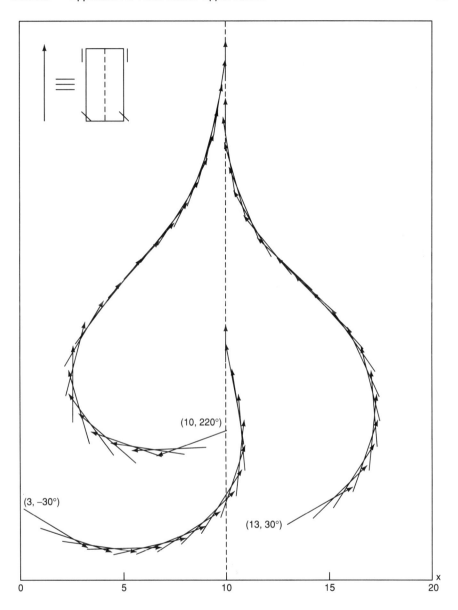

Figure 5.4 Truck trajectories using the neural controller and the numerical-fuzzy controller.

Finally, Step 5 of our numerical-fuzzy method generated a controller for the truck for the three initial states, (x_0, ϕ_0^o) = (3,-30), (10,220), and (13,30), which are the same states used in the simulations of the neural controller. The final trajectories of the truck have no visible difference from Figure 5.4; hence, Figure 5.4 also shows the truck trajectories using the numerical-fuzzy controller.

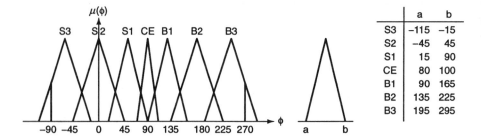

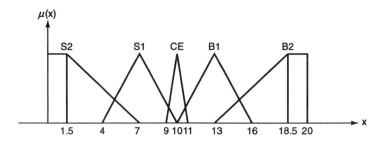

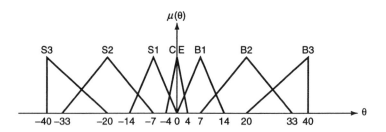

Figure 5.5 Fuzzy membership functions for the truck backer-upper control problem.

		S2	S1	CE	B1	B2
	S3	S2	S3			
	S2	S2	S3	S3	S3	
	S1	B1	S1	S2	S3	S2
φ	CE	B2	B2	CE	S2	S2
	B1	B2	B3	B2	B1	S1
	B2		B3	B3	B3	B2
	B3				B3	B2

(table header row: x)

Figure 5.6 The final fuzzy rule base generated from the numerical data for the truck backer-upper control problem.

Sec. 5.3   Application to Truck Backer-Upper Control   75

TABLE 5.2 Fuzzy rules generated from the
desired input-output pairs of Table 5.1, and the
degrees of these rules.

Fuzzy rules for $t =$	IF x is	IF ϕ is	THEN θ is	Degree
0	S2	S2	S2	1.00
1	S2	S2	S2	0.92
2	S2	S2	S2	0.35
3	S2	S2	S2	0.12
4	S2	S2	S2	0.07
5	S1	S2	S1	0.08
6	S1	S1	S1	0.18
7	S1	S1	S1	0.52
8	S1	S1	S1	0.56
9	S1	S1	S1	0.60
10	CE	S1	S1	0.35
11	CE	S1	S1	0.21
12	CE	S1	CE	0.16
13	CE	CE	CE	0.32
14	CE	CE	CE	0.45
15	CE	CE	CE	0.54
16	CE	CE	CE	0.88
17	CE	CE	CE	0.92
18				
19				
20				

We simulated the neural and numerical-fuzzy controllers for other initial truck positions and observed that the truck trajectories using these two controllers were also almost the same. This is not surprising because both controllers used the same information to construct their control laws.

Example 5.2. In this example we consider the situation where neither linguistic fuzzy rules alone nor desired input-output pairs alone are sufficient to successfully control the truck to the desired position. That is, neither the usual fuzzy controller with limited fuzzy rules nor the usual neural controller can control the truck to the desired position, but a combination of linguistic fuzzy rules and fuzzy rules generated from the desired input-output data pairs is sufficient to successfully control the truck to the desired position.

We consider the case where the beginning part of the information comes from desired input-output pairs, whereas the ending part of the information comes from linguistic rules. To do this we used only the first three pairs of each of the 14 desired sequences and generated fuzzy rules based only on these truncated pairs. The fuzzy rule base generated from these truncated data pairs is the same as Figure 5.6 except that there are no rules in the three center boxes of column CE. The fuzzy rule base of linguistic rules for the ending part was chosen to have only three rules, which are the same as the three center rules of column CE in Figure 5.6.

We simulated the following three cases in which we used: (1) the fuzzy rule base generated from only the truncated data pairs; (2) the fuzzy rule base of selected linguistic rules, and (3) the fuzzy rule base which combined the fuzzy rule bases of the first and second cases. We see that for Case 3 the fuzzy rule base is the same as in Figure 5.6; hence, the truck trajectories for this case must be the same as those using the fuzzy rule base of Figure 5.6. For each of the cases, we simulated the system starting from the following three initial states: $(x_0, \phi_0^o) = (3,-30), (10,220)$, and $(13,30)$. The resulting trajectories for Cases 1, 2, and 3 for the three initial states are shown in Figures 5.7, 5.8, and 5.4, respectively.

We see very clearly from these figures that for Cases 1 and 2 the truck cannot be controlled to the desired position, whereas for Case 3 we successfully controlled the truck to the desired position.

5.4 APPLICATION TO TIME-SERIES PREDICTION

Let $x(k)$ ($k = 1, 2, 3, \ldots$) be a time series. The problem of time-series prediction can be formulated as: given $x(k - n + 1), x(k - n + 2), \ldots, x(k)$, determine $x(k + l)$, where n and l are fixed positive integers. That is, determine a mapping from $[x(k - n + 1), x(k - n + 2), \ldots, x(k)] \in R^n$ to $[x(k + l)] \in R$.

We now use our method in Section 5.2 for the time-series prediction problem. Assuming that $x(1), x(2), \ldots, x(k)$ are given, we can form $k - n$ input-output pairs:

$$[x(k - n), \ldots, x(k - 1); x(k)]$$
$$[x(k - n - 1), \ldots, x(k - 2); x(k - 1)]$$
$$\ldots$$
$$[x(1), \ldots, x(n); x(n + 1)] \tag{5.14}$$

Steps 1–4 of our method in Section 5.2 are used to generate a fuzzy rule base based on the input-output pairs (5.14); then this fuzzy rule base is used to forecast $x(k + p)$ for $p = 1, 2, \ldots$, using the defuzzifying procedure of Step 5 of our method in Section 5.2, where the inputs to the predictor are $x(k + p - n), x(k + p - n + 1), \ldots, x(k + p - 1)$.

Consider that the time series $x(k)$ is generated by the Mackey-Glass chaotic system described by (4.26) with $\tau = 30$. Figure 5.9 shows 1,000 points of this chaotic time series. We chose $n = 9$ and $l = 1$ in our simulation; that is, nine point values in the series were used to predict the value of the next time point. The membership functions for any point are shown in Figure 5.10 for the numerical-fuzzy predictor (later, we use other membership functions). The first 700 points of the series were used as training data, and the final 300 points were used as test

Sec. 5.4 Application to Time-Series Prediction **77**

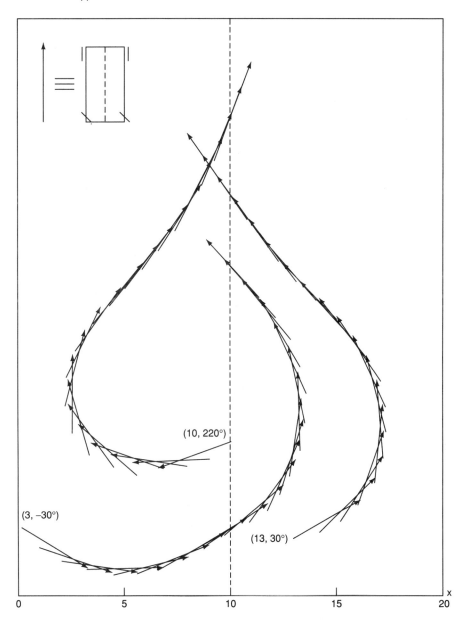

Figure 5.7 Truck trajectories using the fuzzy rules from the truncated data pairs only.

data (for additional cases, see Wang and Mendel [83]). We simulated two cases: (1) 200 training data (from 501 to 700) were used to construct the fuzzy rule base, and (2) 700 training data (from 1 to 700) were used. Figures 5.11 and 5.12 show the results for Cases 1 and 2, respectively.

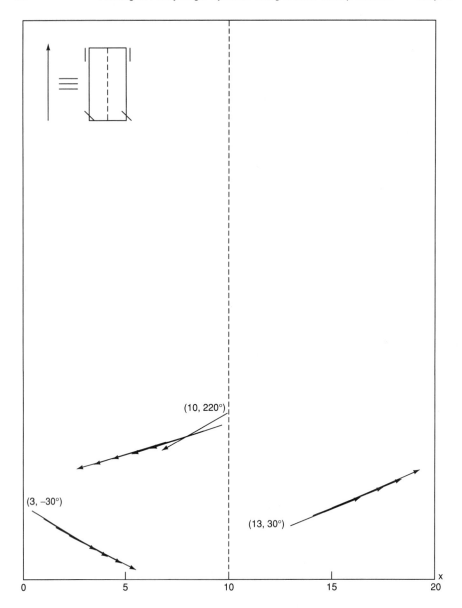

Figure 5.8 Truck trajectories using the selected linguistic rules only.

It is very easy to modify the fuzzy rule base as new data become available. Specifically, when a new data pair becomes available, we create a rule for this data pair and add the new rule to the fuzzy rule base; then the updated (that is, adapted) fuzzy rule base is used to predict the future values. By using this adaptive procedure we use all the available information to predict the next value of the series.

Sec. 5.4   Application to Time-Series Prediction   79

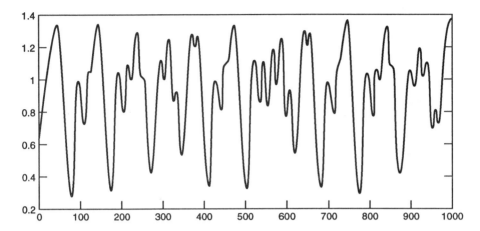

Figure 5.9 A section of the Mackey-Glass chaotic time series.

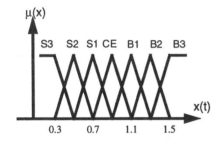

Figure 5.10 The first choice of membership functions for the chaotic time-series prediction problem.

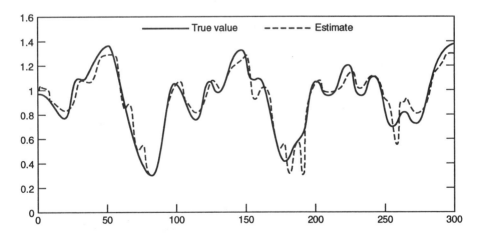

Figure 5.11 Prediction of the chaotic time series from $x(701)$ to $x(1,000)$ using the numerical-fuzzy predictor when 200 training data (from $x(501)$ to $x(700)$) are used.

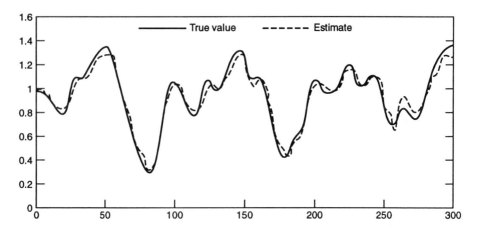

Figure 5.12 Prediction of the chaotic time series from $x(701)$ to $x(1,000)$ using the numerical-fuzzy predictor when 700 training data (from $x(1)$ to $x(700)$) are used.

We simulated this adaptive procedure for the chaotic series of Figure 5.9. We started with the fuzzy rule base generated by the data $x(1)$ to $x(700)$, made a prediction of $x(701)$, then used the true value of $x(701)$ to update the fuzzy rule base, and this updated fuzzy rule base was then used to predict $x(702)$. This adaptive procedure continued until $x(1,000)$. Its results are shown in Figure 5.13. Comparing Figures 5.13 and 5.11, we see that we obtain only a slightly improved prediction.

Finally, we show that prediction can be greatly improved by dividing the domain interval into finer regions. We performed two simulations: one with the membership function shown in Figure 5.14, and the other with the membership

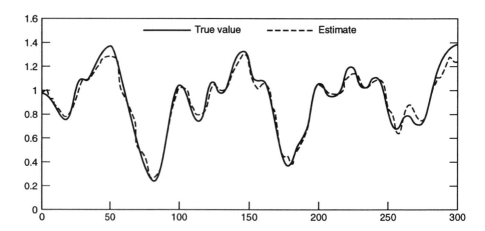

Figure 5.13 Prediction of the chaotic time series from $x(701)$ to $x(1,000)$ using the adaptive fuzzy rule base procedure.

Sec. 5.4 Application to Time-Series Prediction

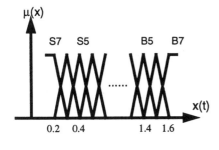

Figure 5.14 The second choice of membership functions for the chaotic time-series prediction problem.

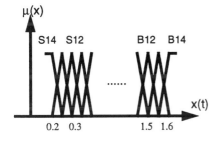

Figure 5.15 The third choice of membership functions for the chaotic time-series prediction problem.

function shown in Figure 5.15. We used the adaptive fuzzy rule base procedure for both simulations. The results are shown in Figures 5.16, and 5.17, where Figure 5.16 (5.17) shows the result corresponding to the membership function of Figure 5.14 (5.15). Comparing Figures 5.13, 5.16 and 5.17 we see very clearly that we obtain better and better results as the domain interval is more finely divided. Figure 5.17 shows that we obtained an almost perfect prediction when we divided the domain interval into 29 regions. Of course, the price paid for doing this is a larger fuzzy rule base.

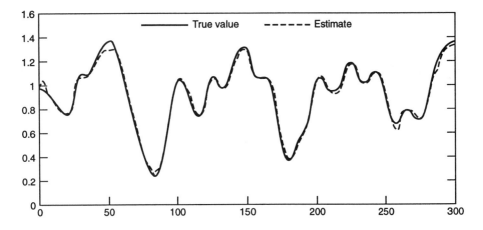

Figure 5.16 Prediction of the chaotic time series from $x(701)$ to $x(1,000)$ using the updating fuzzy rule base procedure with the second choice of membership function.

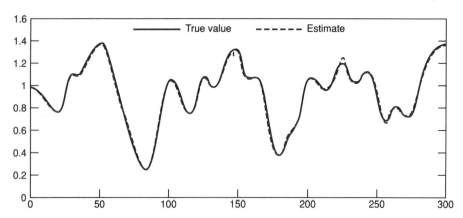

Figure 5.17 Prediction of the chaotic time series from $x(701)$ to $x(1,000)$ using the updating fuzzy rule base procedure with the third choice of membership function.

5.5 CONCLUDING REMARKS

Based on the table-lookup representation of the fuzzy rule base, we developed a general method to generate fuzzy rules from numerical data. The rules generated from numerical data and the rules from human experts were then combined into a common fuzzy rule base based on which the final fuzzy logic system is constructed. We applied this method to a truck backer-upper control problem, and observed that (1) for the same training set (that is, the same given input-output pairs), the final control performance of this method is indistinguishable from that of the pure neural network controller; and (2) in the case where neither numerical data nor linguistic rules contain enough information, both the pure neural and pure fuzzy methods failed to control the truck to the desired position, but this method succeeded. We also applied this method to the Mackey-Glass chaotic time-series prediction problem.

The most important advantage of this method is its simplicity—it just performs a simple one-pass operation on the training data. The price paid for this simplicity is that we have to determine the partitions of the domain intervals and the membership functions in an ad hoc manner.

6

TRAINING OF FUZZY LOGIC SYSTEMS USING NEAREST NEIGHBORHOOD CLUSTERING

6.1 INTRODUCTION

For some practical problems, sample data may be expensive to obtain. For example, a test flight of a new aircraft is very expensive. Although the training algorithms in Chapters 3–5 can be applied to these small-sample problems, they do not take the small-sample factor explicitly into consideration. For these small-sample problems, we may want a fuzzy logic system that is capable of matching all the input-output pairs to any given accuracy.

In this chapter, we first develop such an optimal fuzzy logic system; that is, it is optimal in the sense that it is capable of matching all the input-output pairs in the training set to any given accuracy. The basic idea of this optimal fuzzy logic system is very simple, and in fact has already been shown in Theorem 3.1. That is, we choose the number of rules in the optimal fuzzy logic system equal to the number of input-output pairs in the training set, with one rule responsible for matching one input-output pair.

We then extend the optimal fuzzy logic system to large-sample problems. To do this we first determine clusters of the sample data using the nearest neighborhood clustering algorithm, and then view the clusters as sample data and use the optimal fuzzy logic system to match them. We call this approach an adaptive version of the optimal fuzzy logic system. We also show how to combine linguistic fuzzy IF-THEN rules and numerical input-output pairs using this adaptive fuzzy system.

Finally, we use this adaptive fuzzy system as controllers for nonlinear dynamic systems and simulate the approach for the same examples used in Narendra and Parthasarathy [49] for testing the neural controllers.

6.2 AN OPTIMAL FUZZY LOGIC SYSTEM

Suppose that we are given N input-output pairs $(\underline{x}^l, y^l), l = 1, 2, \ldots, N$, and N is small, say, $N = 20$. Our task is to construct a fuzzy logic system $f(\underline{x})$ which can match all the N pairs to any given accuracy. That is, for any given $\epsilon > 0$, we require that $|f(\underline{x}^l) - y^l| < \epsilon$ for all $l = 1, 2, \ldots, N$.

This optimal fuzzy logic system is constructed as

$$f(\underline{x}) = \frac{\sum_{l=1}^{N} y^l \exp(-\frac{|\underline{x}-\underline{x}^l|^2}{\sigma^2})}{\sum_{l=1}^{N} \exp(-\frac{|\underline{x}-\underline{x}^l|^2}{\sigma^2})} \qquad (6.1)$$

Clearly, the fuzzy logic system (6.1) is in the form of (2.46) (noticing that $\prod_{i=1}^{n} \exp(-(\frac{x_i-x_i^l}{\sigma})^2) = \exp(-\frac{|\underline{x}-\underline{x}^l|^2}{\sigma^2})$). The following theorem shows that by properly choosing the parameter σ, the fuzzy logic system (6.1) can match all the N input-output pairs to any given accuracy.

Theorem 6.1. For arbitrary $\epsilon > 0$, there exists $\sigma^* > 0$ such that the fuzzy logic system (6.1) with $\sigma = \sigma^*$ has the property that

$$|f(\underline{x}^l) - y^l| < \epsilon \qquad (6.2)$$

for all $l = 1, 2, \ldots, N$.

Proof. Viewing $\underline{x}^l$ and y^l as the $\underline{u}(l)$ and $g(\underline{u}(l))$ in Theorem 3.1 and using exactly the same method as the proof of Theorem 3.1, we can prove this theorem. Q.E.D.

We now make two remarks on this optimal fuzzy logic system.

Remark 6.1. The σ is a smoothing parameter: the smaller the σ, the smaller the matching error $|f(\underline{x}^l) - y^l|$, but the less smooth the $f(\underline{x})$ becomes. We know that if $f(\underline{x})$ is not smooth, it may not generalize well for the data points not in the training set. Thus, the σ should be properly chosen to provide a balance between matching and generalization. Because the σ is a one-dimensional parameter, it is usually not difficult to determine an appropriate σ for a practical problem. Sometimes, a few trial-and-error procedures may determine a good σ. As a general rule, large σ can smooth out noisy data, while small σ can make $f(\underline{x})$ as nonlinear as is required to approximate closely the training data.

Remark 6.2. The $f(\underline{x})$ is a general nonlinear regression which provides a smooth interpolation between the observed points $(\underline{x}^l, y^l)$. It is well behaved even for very small σ. In Chapter 7 we show that this $f(\underline{x})$ is identical to the probabilistic general regression of Specht, [67].

6.3 AN ADAPTIVE VERSION OF THE OPTIMAL FUZZY LOGIC SYSTEM

The optimal fuzzy logic system (6.1) uses one rule for one input-output pair in the training set; thus, it is no longer a practical system if the number of input-output pairs in the training set is large. For these large-sample problems, various clustering techniques can be used to group the samples so that a group can be represented by only one rule in the fuzzy logic system. Here we use the simple nearest neighborhood clustering scheme.

An Adaptive Version of the Optimal Fuzzy Logic System

- Starting with the first input-output pair $(\underline{x}^1, y^1)$, establish a cluster center $\underline{x}_0^1$ at $\underline{x}^1$, and set $A^1(1) = y^1$, $B^1(1) = 1$. Select a radius r.
- Suppose that when we consider the kth input-output pair $(\underline{x}^k, y^k)$, $k = 2, 3, \ldots$, there are M clusters with centers at $\underline{x}_0^1, \underline{x}_0^2, \ldots, \underline{x}_0^M$. Compute the distances of $\underline{x}^k$ to these M cluster centers, $|\underline{x}^k - \underline{x}_0^l|$, $l = 1, 2, \ldots, M$, and let the smallest distances be $|\underline{x}^k - \underline{x}_0^{l_k}|$, that is, the nearest cluster to $\underline{x}^k$ is $\underline{x}_0^{l_k}$. Then:

    a) If $|\underline{x}^k - \underline{x}_0^{l_k}| > r$, establish $\underline{x}^k$ as a new cluster center $\underline{x}_0^{M+1} = \underline{x}^k$, set $A^{M+1}(k) = y^k$, $B^{M+1}(k) = 1$, and keep $A^l(k) = A^l(k-1)$, $B^l(k) = B^l(k-1)$ for $l = 1, 2, \ldots, M$.
    b) If $|\underline{x}^k - \underline{x}_0^{l_k}| \leq r$, do the following:

$$A^{l_k}(k) = A^{l_k}(k-1) + y^k \quad (6.3)$$

$$B^{l_k}(k) = B^{l_k}(k-1) + 1 \quad (6.4)$$

    and set

$$A^l(k) = A^l(k-1) \quad (6.5)$$

$$B^l(k) = B^l(k-1) \quad (6.6)$$

    for $l = 1, 2, \ldots, M$ with $l \neq l_k$.

- The adaptive fuzzy system at the kth step is computed as

$$f_k(\underline{x}) = \frac{\sum_{l=1}^{M} A^l(k) exp(-\frac{|\underline{x}-\underline{x}_0^l|^2}{\sigma^2})}{\sum_{l=1}^{M} B^l(k) exp(-\frac{|\underline{x}-\underline{x}_0^l|^2}{\sigma^2})} \quad (6.7)$$

if $\underline{x}^k$ does not establish a new cluster; and, if $\underline{x}^k$ establishes a new cluster, change the M in (6.7) to $M + 1$.

We now make a few remarks on this adaptive fuzzy system.

Remark 6.3. Comparing (6.7) with (6.1) and noticing (6.3)–(6.6), we see that if we replace the $\underline{x}^l$ in (6.1) by the center of the cluster to which the $\underline{x}^l$ belongs, then (6.1) becomes (6.7). This is why we call (6.7) an adaptive version of the optimal fuzzy logic system.

Remark 6.4. The radius r determines the complexity of the adaptive fuzzy system. For smaller radius r, we have more clusters which result in a more sophisticated nonlinear regression at the price of more computation to evaluate it. Because r is a one-dimensional parameter (like σ), we may find an appropriate r for a specific problem by trial and error.

Remark 6.5. Because for each input-output pair a new cluster may be formed, this adaptive fuzzy system performs parameter adaptation as well as structure adaptation in a uniform fashion.

Since the A and B coefficients in (6.7) are determined using recursive equations, it is easy to add a forgetting factor to (6.3)–(6.6). This is desirable if the adaptive fuzzy system is being used to model a system with changing characteristics. For these cases, we replace (6.3) and (6.4) with

$$A^{l_k}(k) = \frac{\tau-1}{\tau} A^{l_k}(k-1) + \frac{1}{\tau} y^k \qquad (6.8)$$

$$B^{l_k}(k) = \frac{\tau-1}{\tau} B^{l_k}(k-1) + \frac{1}{\tau} \qquad (6.9)$$

and replace (6.5) and (6.6) with

$$A^l(k) = \frac{\tau-1}{\tau} A^l(k-1) \qquad (6.10)$$

$$B^l(k) = \frac{\tau-1}{\tau} B^l(k-1) \qquad (6.11)$$

where τ can be considered as a time constant of an exponential decay function. For practical considerations, there should be a lower threshold for B^l so that when sufficient time has elapsed without update for a particular cluster (which results in the B^l to be smaller than the threshold), that cluster would be eliminated.

Finally, we see how to combine linguistic fuzzy IF-THEN rules and numerical input-output pairs using this adaptive fuzzy system. Here we use the second scheme described in Section 1.1, that is, we use linguistic rules and input-output pairs to construct two separate fuzzy logic systems, and the final adaptive fuzzy system is obtained by combining them through weighted average. Specifically, let $f^L(\underline{x})$ be a fuzzy logic system, which can be in any form of (2.40), (2.44), (2.46), or (2.49) constructed from linguistic fuzzy IF-THEN rules. Then the final adaptive fuzzy system is

$$f(\underline{x}) = \alpha f^L(\underline{x}) + (1-\alpha) f_k(\underline{x}) \qquad (6.12)$$

where $f_k(\underline{x})$ is given by (6.7), and $\alpha \in [0, 1]$ is a weighting factor. If there are no linguistic rules, set $\alpha = 0$.

6.4 APPLICATION TO ADAPTIVE CONTROL OF NONLINEAR DYNAMIC SYSTEMS

In this section, we use the adaptive fuzzy system (6.7) as a basic building block of adaptive controllers for nonlinear dynamic systems. We use the same examples as in Narendra and Parthasarathy [49].

Example 6.1. We consider here the problem of controlling the plant discussed in Example 3.2, which is described by the difference equation

$$y(k+1) = g[y(k), y(k-1)] + u(k) \tag{6.13}$$

where the function

$$g[y(k), y(k-1)] = \frac{y(k)y(k-1)[y(k)+2.5]}{1+y^2(k)+y^2(k-1)} \tag{6.14}$$

is assumed to be unknown. The aim of control is to determine a controller $u(k)$ (based on the adaptive fuzzy system) such that the output $y(k)$ of the closed-loop system follows the output $y_m(k)$ of the following reference model:

$$y_m(k+1) = 0.6 y_m(k) + 0.2 y_m(k-1) + r(k) \tag{6.15}$$

where $r(k) = sin(2\pi k/25)$. That is, we want $e(k) = y(k) - y_m(k)$ to converge to zero as k goes to infinity.

If the function $g[*]$ of (6.14) is known, we can construct a controller as follows:

$$u(k) = -g[y(k), y(k-1)] + 0.6y(k) + 0.2y(k-1) + r(k) \tag{6.16}$$

which, when applied to (6.13), results in

$$y(k+1) = 0.6y(k) + 0.2y(k-1) + r(k) \tag{6.17}$$

Combining (6.15) and (6.17), we have

$$e(k+1) = 0.6e(k) + 0.2e(k-1) \tag{6.18}$$

From this, it follows that $lim_{k \to \infty} e(k) = 0$. However, since $g[*]$ is unknown, the controller (6.16) cannot be implemented. To solve this problem, we replace the $g[*]$ in (6.16) by the adaptive fuzzy system (6.7). That is, we use the following controller:

$$u(k) = -f_k[y(k), y(k-1)] + 0.6y(k) + 0.2y(k-1) + r(k) \tag{6.19}$$

where $f_k[*]$ is in the form of (6.7) with $\underline{x} = (y(k), y(k-1))^T$. This results in the nonlinear difference equation

$$y(k+1) = g[y(k), y(k-1)] - f_k[y(k), y(k-1)] + 0.6y(k) + 0.2y(k-1) + r(k) \tag{6.20}$$

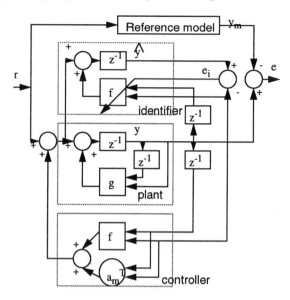

Figure 6.1 Overall adaptive fuzzy control system for Example 6.1.

governing the behavior of the closed-loop system. The overall control system is shown in Figure 6.1. From Figure 6.1 we see that the adaptive fuzzy controller consists of two parts: an identifier and a controller. The identifier uses the adaptive fuzzy system f to approximate the unknown nonlinear function g in the plant, and this f is then copied to the controller.

We simulated the following two cases for this example:

- *Case 1*: The controller in Figure 6.1 was first disconnected and only the identifier was operating to identify the unknown plant. In this identification phase, we chose the input $u(k)$ to be an i.i.d. random signal uniformly distributed in the interval $[-3, 3]$. After the identification procedure was terminated, (6.19) was used to generate the control input; that is, the controller in Figure 6.1 began operating with f copied from the final f in the identifier. Figures 6.2 and 6.3 show the output $y(k)$ of the closed-loop system with this controller together with the reference model output $y_m(k)$ for the cases where the identification procedure was terminated at $k = 100$ and $k = 500$, respectively. In these simulations, we chose $\sigma = 0.3$ and $r = 0.3$. From these simulation results we see that (1) with only 100 steps of training the identifier could produce an accurate model which results in a very good tracking control, and (2) with more steps of training the control performance was improved. In Narendra and Parthasarathy [49], the neural controller achieved similar performance when the identification procedure was carried out for 10^5 steps.

Sec. 6.4 Application to Adaptive Control of Nonlinear Dynamic Systems

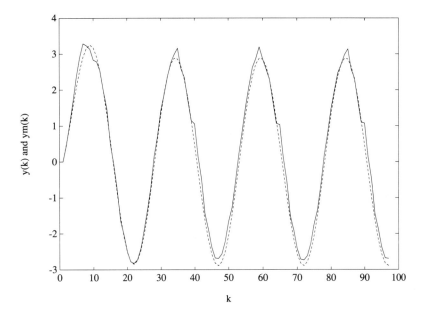

Figure 6.2 The output $y(k)$ (solid line) of the closed-loop system and the reference trajectory $y_m(k)$ (dashed line) for Case 1 in Example 6.1 when the identification procedure was terminated at $k = 100$.

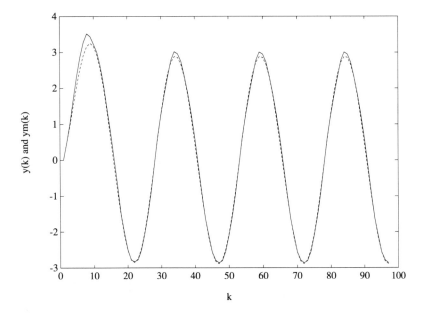

Figure 6.3 The output $y(k)$ (solid line) of the closed-loop system and the reference trajectory $y_m(k)$ (dashed line) for Case 1 in Example 6.1 when the identification procedure was terminated at $k = 500$.

- *Case 2*: The identifier and the controller operated simultaneously (as shown in Figure 6.1) from $k = 0$. We still chose $\sigma = 0.3$ and $r = 0.3$. Figure 6.4 shows $y(k)$ and $y_m(k)$ for this simulation. We see that the control was almost perfect.

Example 6.2. In this example we consider the plant

$$y(k+1) = \frac{5y(k)y(k-1)}{1+y^2(k)+y^2(k-1)+y^2(k-2)} + u(k)$$

$$+0.8u(k-1) \tag{6.21}$$

where the nonlinear function is assumed to be unknown. The aim is to design a controller $u(k)$ such that $y(k)$ will follow the reference model

$$y_m(k+1) = 0.32y_m(k) + 0.64y_m(k-1) - 0.5y_m(k-2)$$

$$+sin(2\pi k/25) \tag{6.22}$$

Using the same idea as in Example 6.1, we choose

$$u(k) = -f_k[y(k), y(k-1), y(k-2)] - 0.8u(k-1) + 0.32y(k) + 0.64y(k-1)$$

$$-0.5y(k-2) + sin(2\pi k/25) \tag{6.23}$$

where $f_k[*]$ is in the form of (6.7). The basic configuration of the overall control scheme is the same as Figure 6.1. Figure 6.5 shows $y(k)$ and $y_m(k)$ when both the

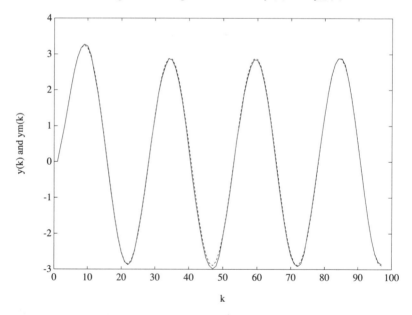

Figure 6.4 The output $y(k)$ (solid line) of the closed-loop system and the reference trajectory $y_m(k)$ (dashed line) for Case 2 in Example 6.1.

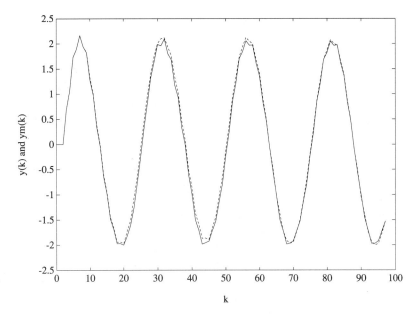

Figure 6.5 The output $y(k)$ (solid line) of the closed-loop system and the reference trajectory $y_m(k)$ (dashed line) for Example 6.2.

identifier and the controller began operating from $k = 0$. We chose $\sigma = 0.3$ and $r = 0.3$ in this simulation. We see, again, that the performance was very good.

6.5 CONCLUDING REMARKS

In this chapter, we first constructed an optimal fuzzy logic system which is capable of matching all the input-output pairs in the training set to arbitrary accuracy. Then we developed an adaptive version of the optimal fuzzy logic system using the nearest neighborhood clustering algorithm. We used this adaptive fuzzy system to construct controllers for nonlinear dynamic systems.

Up to now, we have developed four different training algorithms (in Chapters 3–6) for fuzzy logic systems. Naturally, the reader may ask: How do they compare with each other? If all of them seem applicable to my problem, which method should I try first? We now briefly answer these questions.

The advantages of the back-propagation training algorithm in Chapter 3 are: (1) all the parameters in the fuzzy logic system are updated in a single optimization procedure; therefore, the trained fuzzy logic system may exhibit a very good performance, and (2) it incorporates linguistic information as initial parameters. This is a good approach because linguistic information is adjusted during the adaptation procedure. The disadvantages of the back-propagation training algorithm are:

(1) the system order, M, must be specified a priori by the designer, and (2) it performs a nonlinear search procedure and thus may converge slowly and be trapped at a local minimum.

The advantage of the orthogonal least squares training algorithm in Chapter 4 is that it may produce a simple and well-performing system because it selects the most significant fuzzy basis functions. The disadvantages of it are: (1) it is an off-line algorithm, so all the training data must be available before this algorithm can be used, and (2) it is computationally expensive.

The advantage of the table-lookup training algorithm in Chapter 5 lies in its simplicity. It performs a one-pass operation on the training data and the operation on each data pair is very simple. The disadvantage of it is that the division of the input space and the membership functions of the fuzzy sets defined in the input space must be specified by the designer a priori, and no optimization on them is conducted.

The advantages of the clustering-based training algorithm in Chapter 6 are: (1) it performs a one-pass operation on the training data and is computationally simple, and (2) its performance is well justified because it is based on the optimal fuzzy logic system. The disadvantage of it is that linguistic information cannot be adjusted during the training procedure by using this algorithm. Because linguistic information is usually not precise and sometimes may not be correct, adjustment should be needed in order for the linguistic information to have a more positive impact on the final system.

For the question of which algorithm to use, our first answer is to check the disadvantages of the training algorithms, and delete the algorithms from consideration if their disadvantages cannot be tolerated for the specific problem. If all four algorithms seem applicable to a specific problem, then based on our experience we recommend the following order of consideration: first, try the clustering-based training algorithm; then try either the table-lookup training algorithm or the back-propagation training algorithm; finally, try the orthogonal least squares training algorithm.

7

COMPARISON OF ADAPTIVE FUZZY SYSTEMS WITH ARTIFICIAL NEURAL NETWORKS

7.1 INTRODUCTION

Adaptive fuzzy systems and artificial neural networks share a common objective: to emulate the operation of the human brain. In some sense, artificial neural networks try to emulate the "hardware" of the human brain, whereas adaptive fuzzy systems try to emulate the "software" in the human brain. For example, the multilayer perceptron simulates the basic structure of our brain—a massive connection of a huge number of simple neurons, whereas fuzzy logic systems, as described in Chapter 2, try to simulate our brain from a higher-level input-output point of view. In fact, this kind of distinction is not new. In the early days of computers, two philosophically opposing views of what computers could be emerged and struggled for recognition. One school believed that both minds and digital computers are symbol-manipulating systems. Symbolic logic and programming became the tools of their trade. The opposing school felt that the ultimate goal of computation is better achieved by modeling the brain itself rather than manipulating the mind's symbolic representation of the external world. Although initial demonstrations proved the viability of both approaches, the brain modelers lost some ground when digital computers were successfully used in 1956 by Newell and Simon to solve puzzles and prove theorems. By this time Rosenblatt also successed in building the perceptron and demonstrated the viability of the opposing school. In the decade that followed, the brain modeler school received a severe blow in 1969 when Minsky and Papert [46] claimed that the perceptron approach is fundamentally flawed. The recent renewal of interest in artificial neural networks is prompted by advances in technology as well as the introduction of new learning algorithms. Additionally, it has been recognized that some problems, like recognizing a familiar face, understanding a natural language, and so on, are very difficult to solve using conventional approaches; therefore, it is worth a trial for using artificial neural networks to solve them.

However, both schools ignore an essential factor in the operation of our brain: the fuzziness. It has been nearly 30 years since Zadeh introduced the fuzzy set theory. During this period, the majority of the scientific community did not view the fuzzy theory as a valuable approach. Some prestigious scientists criticized the theory, which resulted in little funding for the research and fewer excellent young students dedicating themselves to this field. Although the fuzzy theory was criticized in the West, it was recognized as a valuable scientific principle in the East. The hard work of the Japanese engineers was rewarded by the impressive successes of fuzzy logic controllers in a variety of customer products and industrial processes. Because of these practical successes, fuzzy theory is beginning to be recognized as a valuable approach in the West. Since the fuzzy research in Japan emphasized applications while the fuzzy research in China concentrated on fuzzy mathematics, we observe the following phenomenon in the field: theory is far behind applications. Developing a comprehensive fuzzy theory from an engineering perspective requires a long-term effort. One question might be where to begin. In this book, we begin by trying to use the ideas in neural network theory to solve fuzzy logic system design problems. Therefore, the aim of this chapter is to compare adaptive fuzzy systems with artificial neural networks.

In Sections 7.2–7.4, we briefly describe three artificial neural networks: multilayer perceptron, radial basis function networks, and probabilistic general regressions, and compare them with adaptive fuzzy systems.

7.2 COMPARISON OF MULTILAYER PERCEPTRON WITH ADAPTIVE FUZZY SYSTEMS

Multilayer perceptrons are feedforward networks with one or more layers of neurons between the input and output layers, where a neuron is a multi-input–single-output element whose output is a nonlinear function of the weighted sum of its inputs. A four-layer perceptron with two hidden layers is shown in Figure 7.1, where the w_{ij}'s are the weights of the neurons (represented by circles), the θ_j's are internal offsets which can be viewed as weights with constant inputs, and the $f(*)$ is usually the sigmoid function:

$$f(x) = \frac{1}{1 + exp(-x)} \tag{7.1}$$

For a given input vector $\underline{x} = (x_1, \ldots, x_n)^T$, the network operates in the forward direction to compute the first hidden layer output $\underline{x}' = (x'_1, \ldots, x'_{n1})^T$, the second hidden layer output $\underline{x}'' = (x''_1, \ldots, x''_{n2})^T$, and the network output $\underline{y} = (y_1, \ldots, y_m)^T$, using the formulas shown in Figure 7.1.

For a given input-output pair $(\underline{x}, \underline{d})$, $\underline{x} \in R^n$ and $\underline{y} \in R^m$, the weights of the multilayer perceptron in Figure 7.1 can be adjusted using the following back-

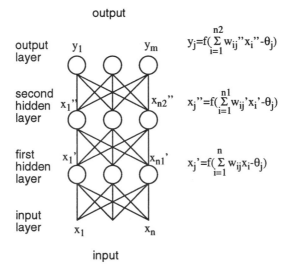

Figure 7.1 A four-layer perceptron with two hidden layers.

propagation training algorithm, which is an iterative gradient algorithm designed to minimize the mean square error between the actual output $\underline{y}$ of the multilayer perceptron and the desired output $\underline{d}$.

The Back-Propagation Training Algorithm for Multilayer Perceptron

Step 1: Initial weights and offsets:

Set all weights and neuron offsets to small random values.

Step 2: Present input and desired output:

Present the $\underline{x}$ in the desired input-output pair $(\underline{x}, \underline{d})$ to the input layer of the multilayer perceptron, and specify the desired network output to be $\underline{d}$. Each input-output pair could be presented for one or more cycles in the back-propagation training procedure. For multiple training samples, the pairs could be presented cyclically until the weights stabilize.

Step 3: Calculate actual output:

Use the sigmoid function (7.1) and the formulas in Figure 7.1 to calculate network output $\underline{y} = (y_1, \ldots, y_m)^T$.

Step 4: Adapt weights:

Use a recursive algorithm starting from the output layer and working backward to the first hidden layer. Adjust weights by

$$w_{ij}(k+1) = w_{ij}(k) + \alpha \delta_j x'_i \qquad (7.2)$$

where $w_{ij}(k)$ is the weight from hidden neuron i or from an input to neuron j at time k, x'_i is either the output of neuron i or is an input, α is a stepsize, and δ_j is an error term for neuron j. If neuron j is an output neuron, then

$$\delta_j = y_j(1 - y_j)(d_j - y_j) \tag{7.3}$$

where d_j is the desired output of neuron j and y_j is the actual output. If neuron j is an internal hidden neuron, then

$$\delta_j = x'_j(1 - x'_j)\sum_l \delta_l w_{jl} \tag{7.4}$$

where l is over all neurons in the layer above neuron j. Internal neuron offsets are adjusted in a similar manner by assuming they are connection weights on links from auxiliary constant inputs.

Step 5: Repeat by going to Step 2.

We now compare the multilayer perceptron (with the preceding back-propagation training algorithm) with the back-propagation adaptive fuzzy system in Chapter 3. They are similar in the following two aspects:

- Their basic operation is the same—forward computation and backward training—and both use iterative gradient algorithms to minimize the mean square error between the actual output and the desired output.
- Both are universal approximators and qualified to solve nonlinear problems.

They are different in the following two aspects:

- The parameters of the adaptive fuzzy system have clear physical meanings and, therefore, very good initial parameter-choosing methods can be developed as shown in Chapter 3. The initial parameters of the multilayer perceptron, as shown earlier, have to be chosen randomly. Also, based on the physical meanings, we can have a feeling of whether a set of parameters is good or not; this feeling is often important in the trial-and-error procedure. On the other hand, by observing a set of parameters (weights) of the multilayer perceptron, it is very difficult to say whether it is good or not.
- Linguistic information can be incorporated into the adaptive fuzzy system in a systematic way, whereas the multilayer perceptron cannot make use of the linguistic information.

7.3 COMPARISON OF RADIAL BASIS FUNCTION NETWORKS WITH ADAPTIVE FUZZY SYSTEMS

The locally-tuned and overlapping receptive field is a well-known structure that has been studied in regions of cerebral cortex, the visual cortex, and so on. Based on the biological receptive fields, Moody and Darken [47] proposed a network

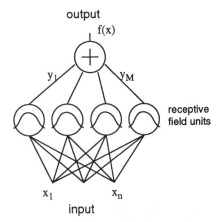

Figure 7.2 Illustration of a radial basis function network.

structure, namely, a radial basis function network that employs local receptive fields to perform function mappings. Figure 7.2 shows the schematic diagram of a radial basis function network with four receptive field units, where the output of the ith receptive field unit is usually a Gaussian function:

$$R_i(\underline{x}) = exp[-\frac{|\underline{x}-\underline{c}_i|^2}{\sigma_i^2}] \tag{7.5}$$

or a logistic function:

$$R_i(\underline{x}) = \frac{1}{1+exp[-\frac{|\underline{x}-\underline{c}_i|^2}{\sigma_i^2}]} \tag{7.6}$$

and the $\underline{c}_i$ and σ_i are parameters. The output of a radial basis function network can be computed in two ways. For the simpler one, as shown in Figure 7.1, the output is a weighted sum of the function values associated with each receptive field:

$$f(\underline{x}) = \sum_{i=1}^{M} \bar{y}_i R_i(\underline{x}) \tag{7.7}$$

where the $\bar{y}_i$ is the strength of the ith receptive field. For the other one, the network produces the normalized response function as the weighted average of the strengths:

$$f(\underline{x}) = \frac{\sum_{i=1}^{M} \bar{y}_i R_i(\underline{x})}{\sum_{i=1}^{M} R_i(\underline{x})} \tag{7.8}$$

To minimize the squared error between the desired output and the network output, we can use training algorithms very similar to the back-propagation algorithm in Chapter 3 and the orthogonal least squares algorithm in Chapter 4. In fact, our training algorithm in Chapter 4 was inspired by the training algorithm for the radial basis function networks developed in Chen, Cowan, and Grant [5]. As another method, Moody and Darken [47] use a self-organizing technique to find the centers $\underline{c}_i$ and widths σ_i of the receptive fields, and then employ the least mean squares algorithm to identify the strengths $\bar{y}_i$.

At this point we may have a feeling that radial basis function networks and adaptive fuzzy systems are the same because (7.8) and (2.46) become identical if we use (7.5) in (7.8) and set $a_i^l = 1$ and $\sigma_i^l = \sigma_i$ in (2.46). However, we would like to emphasize that they are not the same, for the following reasons:

- Functions in the form of (2.46) are just one kind of fuzzy logic systems: that is, they are fuzzy logic systems with a particular choice of fuzzy inference engine, fuzzifier, and defuzzifier. If we take other choices, for example, use the maximum defuzzifier, then the fuzzy logic system will be quite different from the radial basis function networks (7.7) or (7.8). In this sense we could say that the radial basis function network is a special case of fuzzy logic systems.

- The membership functions of the fuzzy logic systems can take many different forms (Gaussian, triangular, trapezoid, logistic, and so on) and can be inhomogeneous (that is, the $\mu_{F_i^l}$'s in (2.40) do not necessarily take the same functional form), whereas the radial basis functions usually take few functional forms and are usually homogeneous. This is due to the different justifications of fuzzy logic systems and radial basis function networks. On one hand, fuzzy logic systems are justified from a fuzzy logic point of view. Therefore, the membership functions can be any functions with range in [0, 1] if human experts find them appropriate to represent their knowledge. On the other hand, the radial basis function networks are based on biological receptive fields. Therefore, it is difficult to justify the use of many different kinds of inhomogeneous basis functions in a single radial basis function network. Technically, it is difficult to specify these different kinds of inhomogeneous basis functions, if we want to use them.

7.4 COMPARISON OF PROBABILISTIC GENERAL REGRESSION WITH ADAPTIVE FUZZY SYSTEMS

First we briefly describe the probabilistic general regression as proposed in Specht [67]. Let $p(\underline{x}, z)$ be the joint probability density function of a vector random variable, $\underline{x} \in R^n$, and a scalar random variable, $z \in R$. The conditional mean of z given $\underline{x}$ (also called the regression of z on $\underline{x}$) is given by

$$E[z|\underline{x}] = \frac{\int_{-\infty}^{\infty} z p(\underline{x}, z) dz}{\int_{-\infty}^{\infty} p(\underline{x}, z) dz} \quad (7.9)$$

Let $(\bar{\underline{x}}^l, \bar{z}^l)$, $l = 1, 2, \ldots, N$, be sample values of the random variables $\underline{x}$ and z; then, a consistent estimator of $p(\underline{x}, z)$ based upon the $(\bar{\underline{x}}^l, \bar{z}^l)$'s, which was proposed by Parzen [54] and was shown in Specht [68] to be a good choice for estimating $p(\underline{x}, z)$, is given by

$$\hat{p}(\underline{x}, z) = \frac{1}{(2\pi)^{(n+1)/2}\sigma^{(n+1)}} \frac{1}{N} \sum_{l=1}^{N} exp[-\frac{(\underline{x} - \underline{\bar{x}}^l)^T(\underline{x} - \underline{\bar{x}}^l)}{2\sigma^2}]exp[-\frac{(z - \bar{z}^l)^2}{2\sigma^2}]$$
(7.10)

Substituting (7.10) into (7.9) and performing the integration yields the following:

$$f(\underline{x}) = \hat{E}(z|\underline{x}) = \frac{\sum_{l=1}^{N} \bar{z}^l exp[-\frac{(\underline{x}-\underline{\bar{x}}^l)^T(\underline{x}-\underline{\bar{x}}^l)}{2\sigma^2}]}{\sum_{l=1}^{N} exp[-\frac{(\underline{x}-\underline{\bar{x}}^l)^T(\underline{x}-\underline{\bar{x}}^l)}{2\sigma^2}]}$$
(7.11)

which is the probabilistic general regression. Analysis of this probabilistic general regression can be found in Specht [67]; it has nice probabilistic properties, like consistency, and so on.

Comparing (7.11) with (2.46), we see that they are almost the same. More specifically, we have the following:

Observation. The probabilistic general regression (7.11) is a special case of fuzzy logic systems. Specifically, within the subset of fuzzy logic systems in Lemma 2.3, if we choose the parameters (in (2.46)): $M = N$, $a_i^l = 1$, $\sigma_i^l = \sigma$ for all $i = 1, 2, \ldots, n$ and $l = 1, 2, \ldots, N$, $\bar{x}_i^l = $ the ith element of the sample vector $\underline{\bar{x}}^l$, $\bar{y}^l = \bar{z}^l$ in the samples, then the fuzzy logic system (2.46) becomes the probabilistic general regression (7.11) (note that $\prod_{i=1}^{n} exp[-\frac{1}{2}(\frac{x_i-\bar{x}_i^l}{\sigma})^2] = exp[-\frac{1}{2\sigma^2}\sum_{i=1}^{n}(x_i - \bar{x}_i^l)^2] = exp[-\frac{(\underline{x}-\underline{\bar{x}}^l)^T(\underline{x}-\underline{\bar{x}}^l)}{2\sigma^2}]$).

We now extend this Observation by making a few remarks:

- Fuzzy logic systems are constructed and justified based on fuzzy logic; very little is known about their statistical properties when they are used in a random environment. From this Observation we see that although there is no statistical consideration when constructing such a fuzzy logic system, the resulting fuzzy logic system turns out to be "optimal" from a statistical point of view.

- From Section 2.6 we see that the fuzzy logic systems in Lemma 2.3 are only a very small subset of fuzzy logic systems; and from this Observation we see that the probabilistic general regression is a special case in this small subset. Therefore, there remains a huge number of fuzzy logic systems for which a statistical study is needed. Up to now, it seems that the successful application of fuzzy logic systems is to the control of industrial processes where random noise always exists. Therefore, knowing the statistical properties of various kinds of fuzzy logic systems is important. There are many fuzzy logic systems that are different from the probabilistic general regression, for example, the fuzzy systems with min inference and triangular membership function, which have found successful practical applications. We hope that these fuzzy logic systems also have nice statistical properties.

- Although the Observation shows the similarity between fuzzy logic systems and the probabilistic general regression, they are different from many fundamental points of view. For example, fuzzy logic systems are constructed from fuzzy IF-THEN rules, whereas the probabilistic general regression is constructed from sample data pairs. It is not difficult to generate reasonable fuzzy IF-THEN rules based on the sample pairs, for example, using the methods in Chapters 3–6. However, it is difficult to generate reasonable data pairs based on fuzzy IF-THEN rules because fuzzy IF-THEN rules characterize fuzzy relationships between input and output so that if we want to generate data pairs based on them, we have to use fuzzification, defuzzification, and inference strategies which will have the same influence on the generated data pairs as the original fuzzy IF-THEN rules.
- Perhaps the most fundamental difference between fuzzy logic systems and the probabilistic general regression is that fuzzy logic systems provide a systematic framework to combine linguistic information in natural language (in the form of fuzzy IF-THEN rules) and measured numerical information (in the form of sample data pairs) in a uniform fashion, whereas the probabilistic general regression can only make use of the numerical information.

7.5 CONCLUDING REMARKS

From this chapter we may conclude the following:

- Although the methodologies of artificial neural networks and adaptive fuzzy systems are different, the final functional forms of them in many cases are quite similar and even exactly the same. This observation has benefited the fuzzy system research—most training algorithms in Chapters 3–6 were inspired by the corresponding training algorithms in the neural network field.
- Adaptive fuzzy systems constitute a much larger functional space than artificial neural networks, that is, in many cases adaptive fuzzy systems include artificial neural networks as special cases. There remains a large number of fuzzy logic systems for which training algorithms can be developed and the performance of which needs to be analyzed.
- The most fundamental difference between adaptive fuzzy systems and artificial neural networks is that the former takes linguistic information explicitly into consideration and makes use of it in a systematic way, whereas the latter does not.

Finally, we make some further comments. Bayesian statisticians believe that probability is sufficient to represent any kind of uncertainty; therefore, human experts should express their knowledge about uncertain events in terms of (conditional) probabilities [Lindley, 35]. However, there is so much human knowledge about uncertain events that is expressed in natural languages, and it is far from easy to "translate" it into probabilistic terms. Fuzzy system researchers have found fuzzy logic systems to be a very *friendly* tool to represent linguistic information in natural languages, and have proven the usefulness of fuzzy logic systems by successfully applying them to a variety of practical problems. From this chapter we see that the superior performance of fuzzy logic systems is not a surprise because fuzzy logic systems include the "optimal" probabilistic general regression and radial basis function networks as special cases. Therefore, by carefully searching the whole space of fuzzy logic systems, the performance of the resulting fuzzy logic system should be no worse than the probabilistic general regression and the radial basis function network.

As for the debate on fuzziness versus probablity, we think, from an engineering point of view, that the important point is *not* whether membership functions are the same as probability distribution functions; *the point is how these functions are used.* More specifically, a probability distribution is used to characterize a collection of data (recall that probability is a measure over a σ-field of subsets of a sample space), whereas a membership function is used to characterize a word in natural languages. Consequently, although the starting representations of the two theories (that is, the membership functions and the probability distributions) may be similar, the directions along which the theories are further developed and used are quite different. Probability theory, on one hand, concentrates, for example, on the statistical analysis of a large amount of data and on the modeling of random noise; fuzzy theory, on the other hand, concentrates on the representation and effective use of human knowledge.

We conclude this chapter by quoting what L. Ljung [38, p. 27] wrote: "... An issue in system identification is to have an open mind about nonlinear black-box structures; to try out the above [Volterra series, neural nets, general regressions, etc.] along with many other ideas, like ..."

8

STABLE INDIRECT ADAPTIVE FUZZY CONTROL OF NONLINEAR SYSTEMS

8.1 INTRODUCTION

When the fuzzy logic systems described in Chapter 2 are used as controllers, they are called *fuzzy logic controllers*. Fuzzy control is by far the most successful application of fuzzy sets and systems theory to practical problems. The present interest in fuzzy theory is largely due to the successful applications of the fuzzy logic controllers to a variety of consumer products and industrial systems [Takagi, 71]. Why has fuzzy control been so successful?

8.1.1 Why Fuzzy Control?

It may be helpful to divide the reasons for fuzzy control into two categories: theoretical and practical reasons. Theoretical reasons for fuzzy control are:

- As a general rule, a good engineering approach should be able to make effective use of all the available information. If the mathematical model of a system is too hard to obtain (this is true for many practical systems), then as we state throughout this book, the most important information comes from two sources: (1) sensors which provide numerical measurements of key variables, and (2) human experts who provide linguistic descriptions about the system and control instructions. Fuzzy controllers, by design, provide a systematic and efficient framework to incorporate linguistic fuzzy information from human experts. Conventional controllers, however, cannot incorporate the linguistic fuzzy information into their designs. If in some situations the most important information comes from human experts, then fuzzy control is the best choice.

Sec. 8.1 Introduction

- Fuzzy control is a model-free approach; that is, it does not require a mathematical model of the system under control. Control engineers are now facing more and more complex systems, and the mathematical models of these systems are more and more difficult to obtain. Thus, model-free approaches become more and more important in control engineering. Conventional control also has some model-free approaches, such as nonlinear adaptive control and PID control. Fuzzy control provides yet another model-free approach.

- Fuzzy control provides nonlinear controllers, which are well justified due to the Universal Approximation Theorem in Section 2.7; that is, these fuzzy logic controllers are general enough to perform any nonlinear control actions. Therefore, by carefully choosing the parameters of the fuzzy controllers, it is always possible to design a fuzzy controller that is suitable for the nonlinear system under control.

In the preceding theoretical reasons, fuzzy control is viewed as a theory and is evaluated from a theoretical point of view. We know that theory and practice have a different emphasis. For example, theory emphasizes generality and rigor, while practice emphasizes applicability to particular problems. Practical reasons for fuzzy control are:

- Fuzzy control is easy to understand. Because fuzzy control emulates human control strategy, its principle is easy to understand for noncontrol specialists. During the last two decades, conventional control theory has been using more and more advanced mathematical tools. This is needed in order to solve difficult problems in a rigorous fashion; however, this also results in fewer and fewer practical engineers who can understand the theory. Therefore, practical engineers who are on the front line of designing consumer products tend to use the approaches which are simple and easy to understand. Fuzzy control is just such an approach.

- Fuzzy control is simple to implement. From Chapters 2 and 3 we see that the fuzzy logic systems, which are the heart of fuzzy control, admit a high degree of parallel implementation. Many fuzzy VLSI chips have been developed [Togai and Watanabe, 75; Togai and Chiu, 76] which make the implementation of fuzzy controllers simple and fast.

- Fuzzy control is cheap to develop. From a practical point of view, the developing cost is one of the most important criterion for a successful product. Because fuzzy control is easy to understand, the time to learn the approach is short; that is, the "software cost" is low. Also, because fuzzy control is simple to implement, the "hardware cost" is also low. Additionally, there are software tools available for designing fuzzy controllers. Thus, fuzzy control is an approach that has a high performance/cost ratio.

8.1.2 Why Adaptive Fuzzy Control?

Fuzzy controllers are supposed to work in situations where there is a large uncertainty or unknown variation in plant parameters and structures. Generally, the basic objective of adaptive control is to maintain consistent performance of a system in the presence of these uncertainties. Therefore, advanced fuzzy control should be adaptive.

What is adaptive fuzzy control? Roughly speaking, if a controller is constructed from adaptive fuzzy systems (recall that an adaptive fuzzy system is a fuzzy logic system equipped with a training (adaptation) algorithm), it is called an adaptive fuzzy controller. An adaptive fuzzy controller can be a single adaptive fuzzy system, or it can be constructed from several adaptive fuzzy systems.

How does an adaptive fuzzy controller compare with a conventional adaptive controller? The most important advantage of adaptive fuzzy control over conventional adaptive control is that adaptive fuzzy controllers are capable of incorporating linguistic fuzzy information from human operators, whereas conventional adaptive controllers cannot. This is especially important for the systems with a high degree of uncertainty, such as chemical processes, aircraft, and so on, because although these systems are difficult to control from a control theoretical point of view, they are often successfully controlled by human operators. How can human operators successfully control such a complex system without a mathematical model in their minds? If we ask the human operators what their control strategies are, they may just tell us a few control rules in fuzzy terms and some linguistic descriptions about the behavior of the system under various conditions which are, of course, also in fuzzy terms. Although these fuzzy control rules and descriptions are not precise and may not be sufficient for constructing a successful controller, they provide very important information about how to control the system and how the system behaves. Adaptive fuzzy control provides a tool for making use of the fuzzy information in a systematic and efficient manner.

How do we classify the adaptive fuzzy controllers? We classify the adaptive fuzzy controllers according to two criteria: (1) whether the adaptive fuzzy controller can incorporate fuzzy control rules or fuzzy descriptions about the system, and (2) whether the fuzzy logic systems in the adaptive fuzzy controller are linear or nonlinear in their adjustable parameters. We detail these classifications in the next two subsections.

8.1.3 Direct and Indirect Adaptive Fuzzy Control

In the conventional adaptive control literature, adaptive controllers are classified into two categories [Narendra and Parthasarathy, 49]: direct and indirect adaptive controllers. In direct adaptive control, the parameters of the controller are directly adjusted to reduce some norm of the output error between the plant and the reference model. In indirect adaptive control, the parameters of the plant are estimated and the controller is chosen assuming that the estimated parameters represent the true values of the plant parameters.

In fuzzy control, linguistic information from human experts can be classified into two categories:

- Fuzzy control rules which state in what situations what control actions should be taken (for example, we often use the following fuzzy IF-THEN rule to drive a car: "IF the speed is slow, THEN apply more force to the accelerator," where "slow" and "more" are labels of fuzzy sets).
- Fuzzy IF-THEN rules which describe the behavior of the unknown plant (for example, we can describe the behavior of a car using the fuzzy IF-THEN rule: "IF you apply more force to the accelerator, THEN the speed of the car will increase," where "more" and "increase" are labels of fuzzy sets).

Interestingly enough, adaptive fuzzy controllers which make use of these two classes of linguistic information correspond to the direct and indirect adaptive control schemes, respectively. More specifically, direct adaptive fuzzy controllers use fuzzy logic systems as controllers; therefore, linguistic fuzzy control rules can be directly incorporated into the controllers. On the other hand, indirect adaptive fuzzy controllers use fuzzy logic systems to model the plant and construct the controllers assuming that the fuzzy logic systems represent the true plant; therefore, fuzzy IF-THEN rules describing the plant can be directly incorporated into the indirect adaptive fuzzy controller. Formally, we have the following definition:

- If an adaptive fuzzy controller uses fuzzy logic systems as controllers, it is called a *direct adaptive fuzzy controller*. A direct adaptive fuzzy controller can incorporate fuzzy control rules directly into itself.
- If an adaptive fuzzy controller uses fuzzy logic systems as a model of the plant, it is called an *indirect adaptive fuzzy controller*. An indirect adaptive fuzzy controller can incorporate fuzzy descriptions of the plant (in terms of fuzzy IF-THEN rules) directly into itself.

8.1.4 First and Second Types of Adaptive Fuzzy Control

From Chapters 3 and 4 we see that the training algorithms for the fuzzy logic system can be quite different depending upon whether the fuzzy logic system is linear or nonlinear in its adjustable parameters. If the fuzzy logic system is linear in its adjustable parameters, then it is easier to find an optimal fuzzy logic system. However, because the searching space is limited to the fuzzy logic systems which are linear in their adjustable parameters, the optimal fuzzy logic system in this searching space may not be good enough. On the other hand, if the fuzzy logic system is nonlinear in its adjustable parameters, then it is more difficult to find an optimal fuzzy logic system. However, if such an optimal fuzzy logic system can be found, its performance should be good because the searching space is large. Thus, the performance, complexity, and adaptive law of an adaptive fuzzy controller can be quite different depending upon whether the fuzzy logic systems in the adaptive fuzzy controller are linear or nonlinear in their adjustable parameters. Therefore, we classify adaptive fuzzy controllers into two types:

- If the fuzzy logic systems used in an adaptive fuzzy controller are linear in their adjustable parameters, this adaptive fuzzy controller is called a *first-type adaptive fuzzy controller*.
- If the fuzzy logic systems used in an adaptive fuzzy controller are nonlinear in their adjustable parameters, this adaptive fuzzy controller is called a *second-type adaptive fuzzy controller*.

Notice that both first and second types of adaptive fuzzy controllers are nonlinear adaptive controllers. We now specify the formula of the fuzzy logic systems used in the first and second types of adaptive fuzzy controllers.

In the first-type adaptive fuzzy controller, we use the following fuzzy logic system:

$$f(\underline{x}) = \sum_{l=1}^{M} \theta_l \xi_l(\underline{x})$$
$$= \underline{\theta}^T \underline{\xi}(\underline{x}) \tag{8.1}$$

where $\underline{\theta} = (\theta_1, \ldots, \theta_M)^T$, $\underline{\xi}(\underline{x}) = (\xi_1(\underline{x}), \ldots, \xi_M(\underline{x}))^T$, $\xi_l(\underline{x})$ is the fuzzy basis function (Chapter 4) defined by

$$\xi_l(\underline{x}) = \frac{\prod_{i=1}^{n} \mu_{F_i^l}(x_i)}{\sum_{l=1}^{M} \prod_{i=1}^{n} \mu_{F_i^l}(x_i)} \tag{8.2}$$

θ_l are adjustable parameters, and $\mu_{F_i^l}$ are *given* membership functions. Clearly, (8.1) is equivalent to (2.40) assuming that $\mu_{F_i^l}$ are given; that is, $\mu_{F_i^l}$ will not change during the adaptation (training) procedure. $\mu_{F_i^l}$ can be Gaussian, triangular, or any other type of membership functions.

In the second-type adaptive fuzzy controller, we use the following fuzzy logic system:

$$f(\underline{x}) = \frac{\sum_{l=1}^{M} \bar{y}^l [\prod_{i=1}^{n} exp(-(\frac{x_i - \bar{x}_i^l}{\sigma_i^l})^2)]}{\sum_{l=1}^{M} [\prod_{i=1}^{n} exp(-(\frac{x_i - \bar{x}_i^l}{\sigma_i^l})^2)]} \tag{8.3}$$

where $\bar{y}^l$, $\bar{x}_i^l$, and σ_i^l are adjustable parameters. Clearly, (8.3) is (2.46) with $a_i^l = 1$.

8.2 A CONSTRUCTIVE LYAPUNOV SYNTHESIS APPROACH TO INDIRECT ADAPTIVE FUZZY CONTROLLER DESIGN

In this section, we first set up the control objectives, and then show in a constructive manner how to develop indirect adaptive controllers based on the fuzzy logic systems to achieve these control objectives.

8.2.1 Control Objectives

Consider the nth-order nonlinear systems of the form

$$\dot{x}_1 = x_2$$
$$\dot{x}_2 = x_3$$
$$\cdots$$
$$\dot{x}_n = f(x_1, \ldots, x_n) + g(x_1, \ldots, x_n)u \quad (8.4)$$
$$y = x_1$$

or equivalently of the form

$$x^{(n)} = f(x, \dot{x}, \ldots, x^{(n-1)}) + g(x, \dot{x}, \ldots, x^{(n-1)})u, \quad y = x \quad (8.5)$$

where f and g are *unknown* continuous functions, $u \in R$ and $y \in R$ are the input and output of the system, respectively, and $\underline{x} = (x_1, x_2, \ldots, x_n)^T = (x, \dot{x}, \ldots, x^{(n-1)})^T \in R^n$ is the state vector of the system which is assumed to be available for measurement. In order for (8.5) to be controllable, we require that $g(\underline{x}) \neq 0$ for $\underline{x}$ in certain controllability region $U_c \subset R^n$; since $g(\underline{x})$ is continuous, without loss of generality we assume that $g(\underline{x}) > 0$ for $\underline{x} \in U_c$. In terms of the nonlinear control literature [Isidori, 21; Slotine and Li, 66], these systems are in normal form and have the relative degree equal to n. The control objective is to force y to follow a given bounded reference signal $y_m(t)$, under the constraint that all signals involved must be bounded. More specifically, we have:

Control Objectives. Determine a feedback control $u = u(\underline{x}|\underline{\theta})$ (based on fuzzy logic systems) and an adaptive law for adjusting the parameter vector $\underline{\theta}$ such that:

1. The closed-loop system must be globally stable in the sense that all variables, $\underline{x}(t), \underline{\theta}(t)$, and $u(\underline{x}|\underline{\theta})$ must be uniformly bounded; that is, $|\underline{x}(t)| \leq M_x < \infty$, $|\underline{\theta}(t)| \leq M_\theta < \infty$ and $|u(\underline{x}|\underline{\theta})| \leq M_u < \infty$ for all $t \geq 0$, where M_x, M_θ, and M_u are design parameters specified by the designer.
2. The tracking error, $e = y_m - y$, should be as small as possible under the constraints in the previous objective.

In the rest of this section, we show the basic ideas of how to construct indirect adaptive fuzzy controllers to achieve these control objectives.

To begin, let $\underline{e} = (e, \dot{e}, \ldots, e^{(n-1)})^T$ and $\underline{k} = (k_n, \ldots, k_1)^T \in R^n$ be such that all roots of the polynomial $h(s) = s^n + k_1 s^{n-1} + \cdots + k_n$ are in the open left half-plane. If the functions f and g are known, then the control law

$$u = \frac{1}{g(\underline{x})}[-f(\underline{x}) + y_m^{(n)} + \underline{k}^T \underline{e}] \tag{8.6}$$

applied to (8.5) results in

$$e^{(n)} + k_1 e^{(n-1)} + \cdots + k_n e = 0 \tag{8.7}$$

which implies that $lim_{t \to \infty} e(t) = 0$—a main objective of control. However, f and g are unknown—what should we do?

8.2.2 Certainty Equivalent Controller

We replace f and g in (8.6) by the fuzzy logic systems $\hat{f}(\underline{x}|\underline{\theta}_f)$ and $\hat{g}(\underline{x}|\underline{\theta}_g)$, respectively, which are in the form of either (8.1) or (8.3). The resulting control law

$$u_c = \frac{1}{\hat{g}(\underline{x}|\underline{\theta}_g)}[-\hat{f}(\underline{x}|\underline{\theta}_f) + y_m^{(n)} + \underline{k}^T \underline{e}] \tag{8.8}$$

is the so-called *certainty equivalent controller* [Sastry and Bodson, 63] in the adaptive control literature. Applying (8.8) to (8.5) and after straightforward manipulation, we obtain the error equation

$$e^{(n)} = -\underline{k}^T \underline{e} + [\hat{f}(\underline{x}|\underline{\theta}_f) - f(\underline{x})] + [\hat{g}(\underline{x}|\underline{\theta}_g) - g(\underline{x})]u_c \tag{8.9}$$

or equivalently

$$\dot{\underline{e}} = \Lambda_c \underline{e} + \underline{b}_c[(\hat{f}(\underline{x}|\underline{\theta}_f) - f(\underline{x})) + (\hat{g}(\underline{x}|\underline{\theta}_g) - g(\underline{x}))u_c] \tag{8.10}$$

where

$$\Lambda_c = \begin{bmatrix} 0 & 1 & 0 & 0 & \cdots & 0 & 0 \\ 0 & 0 & 1 & 0 & \cdots & 0 & 0 \\ \cdots & \cdots & \cdots & \cdots & \cdots & \cdots & \cdots \\ 0 & 0 & 0 & 0 & \cdots & 0 & 1 \\ -k_n & -k_{n-1} & \cdots & \cdots & \cdots & \cdots & -k_1 \end{bmatrix}, \underline{b}_c = \begin{bmatrix} 0 \\ \cdots \\ 0 \\ 1 \end{bmatrix} \tag{8.11}$$

Since Λ_c is a stable matrix ($|sI - \Lambda_c| = s^{(n)} + k_1 s^{(n-1)} + \cdots + k_n$ which is stable), we know that there exists a unique positive definite symmetric $n \times n$ matrix P which satisfies the Lyapunov equation [Slotine and Li, 66]:

$$\Lambda_c^T P + P \Lambda_c = -Q \tag{8.12}$$

where Q is an arbitrary $n \times n$ positive definite matrix. Let $V_e = \frac{1}{2}\underline{e}^T P \underline{e}$, then using (8.10) and (8.12) we have

$$\dot{V}_e = \frac{1}{2}\underline{\dot{e}}^T P \underline{e} + \frac{1}{2}\underline{e}^T P \underline{\dot{e}}$$

$$= -\frac{1}{2}\underline{e}^T Q \underline{e} + \underline{e}^T P \underline{b}_c [(\hat{f}(\underline{x}|\underline{\theta}_f) - f(\underline{x})) + (\hat{g}(\underline{x}|\underline{\theta}_g) - g(\underline{x}))u_c] \quad (8.13)$$

In order for $x_i = y_m^{(i-1)} - e^{(i-1)}$ to be bounded, we require that V_e must be bounded, which means we require that $\dot{V}_e \leq 0$ when V_e is greater than a large constant $\bar{V}$. However, from (8.13) we see that it is very difficult to design the u_c such that the last term of (8.13) is less than zero. How can this problem be solved?

8.2.3 Supervisory Control

We solve this problem by appending another control term, u_s, to the u_c. That is, the final control is

$$u = u_c + u_s \quad (8.14)$$

This additional control term u_s is called a *supervisory control* for the reasons given at the end of this subsection. The purpose of this supervisory control u_s is to force $\dot{V}_e \leq 0$ when $V_e > \bar{V}$. Substituting (8.14) into (8.5) and using the same manipulation for obtaining (8.10), we have the new error equation:

$$\underline{\dot{e}} = \Lambda_c \underline{e} + \underline{b}_c [(\hat{f}(\underline{x}|\underline{\theta}_f) - f(\underline{x})) + (\hat{g}(\underline{x}|\underline{\theta}_g) - g(\underline{x}))u_c - g(\underline{x})u_s] \quad (8.15)$$

Using (8.15) and (8.12), we have

$$\dot{V}_e = -\frac{1}{2}\underline{e}^T Q \underline{e} + \underline{e}^T P \underline{b}_c [(\hat{f}(\underline{x}|\underline{\theta}_f) - f(\underline{x})) + (\hat{g}(\underline{x}|\underline{\theta}_g) - g(\underline{x}))u_c - g(\underline{x})u_s]$$

$$\leq -\frac{1}{2}\underline{e}^T Q \underline{e} + |\underline{e}^T P \underline{b}_c|[|\hat{f}(\underline{x}|\underline{\theta}_f)| + |f(\underline{x})| + |\hat{g}(\underline{x}|\underline{\theta}_g)u_c| + |g(\underline{x})u_c|]$$

$$- \underline{e}^T P \underline{b}_c g(\underline{x}) u_s \quad (8.16)$$

In order to design the u_s such that the right-hand side of (8.16) is nonpositive, we need to know the bounds of f and g. That is, we have to make the following assumption.

Assumption 8.1. We can determine functions $f^U(\underline{x}), g^U(\underline{x})$ and $g_L(\underline{x})$ such that $|f(\underline{x})| \leq f^U(\underline{x})$ and $g_L(\underline{x}) \leq g(\underline{x}) \leq g^U(\underline{x})$ for $\underline{x} \in U_c$, where $f^U(\underline{x}) < \infty$, $g^U(\underline{x}) < \infty$, and $g_L(\underline{x}) > 0$ for $\underline{x} \in U_c$.

Because of Assumption 8.1, the plant (8.5) can be viewed as "poorly-understood," but not "totally unknown." Note that in Assumption 8.1 we require to know the state-dependent bounds of f and g, which is less restrictive than requiring fixed bounds for all $\underline{x} \in U_c$.

Based on f^U, g^U, and g_L, and by observing (8.16), we choose the supervisory control u_s as

$$u_s = I_1^* sgn(\underline{e}^T P\underline{b}_c) \frac{1}{g_L(\underline{x})}[|\hat{f}(\underline{x}|\underline{\theta}_f)| + f^U(\underline{x}) + |\hat{g}(\underline{x}|\underline{\theta}_g)u_c| + |g^U(\underline{x})u_c|] \tag{8.17}$$

where $I_1^* = 1$ if $V_e > \bar{V}$ (which is a constant specified by the designer), $I_1^* = 0$ if $V_e \leq \bar{V}$, and $sgn(y) = 1(-1)$ if $y \geq 0$ (< 0). Substituting (8.17) into (8.16) and considering the case $V_e > \bar{V}$, we have

$$\dot{V}_e \leq -\frac{1}{2}\underline{e}^T Q\underline{e} + |\underline{e}^T P\underline{b}_c|[|\hat{f}| + |f| + |\hat{g}u_c| + |gu_c|$$
$$- \frac{g}{g_L}(|\hat{f}| + f^U + |\hat{g}u_c| + |g^U u_c|)]$$
$$\leq -\frac{1}{2}\underline{e}^T Q\underline{e} \leq 0 \tag{8.18}$$

In summary, using the control (8.14) with u_c given by (8.8) and u_s given by (8.17), we can guarantee that $V_e \leq \bar{V} < \infty$. Since P is positive definite, the boundedness of V_e implies the boundedness of $\underline{e}$, which in turn implies the boundedness of $\underline{x}$. Note that all the quantities in the right-hand sides of (8.8) and (8.17) are known or available for measurement. Therefore, the control law (8.14) can be implemented.

From (8.17) we see that the u_s is nonzero only when the error function V_e is greater than a positive constant $\bar{V}$. That is, if the closed-loop system with the fuzzy controller u_c of (8.8) is well behaved in the sense that the error is not big (that is, $V_e \leq \bar{V}$), then the supervisory control u_s is zero. On the other hand, if the system tends to be unstable (that is, $V_e > \bar{V}$), then the supervisory control u_s begins to operate to force $V_e \leq \bar{V}$. In this way, the control u_s is like a *supervisor*. This is why we call the u_s a supervisory control.

Our next task, in this constructive route, is to replace $\hat{f}$ and $\hat{g}$ by specific formulas of fuzzy logic systems of (8.1) or (8.3), and to develop an adaptive law to adjust the parameters in the fuzzy logic systems for the purpose of forcing the tracking error to converge to zero.

8.2.4 Adaptive Law

First, define

$$\underline{\theta}_f^* = argmin_{\underline{\theta}_f \in \Omega_f}[sup_{\underline{x} \in U_c}|\hat{f}(\underline{x}|\underline{\theta}_f) - f(\underline{x})|] \tag{8.19}$$

$$\underline{\theta}_g^* = argmin_{\underline{\theta}_g \in \Omega_g}[sup_{\underline{x} \in U_c}|\hat{g}(\underline{x}|\underline{\theta}_g) - g(\underline{x})|] \tag{8.20}$$

where Ω_f and Ω_g are constraint sets for $\underline{\theta}_f$ and $\underline{\theta}_g$, respectively, specified by the designer. For Ω_f, we require that $\underline{\theta}_f$ is bounded, and for the fuzzy logic system (8.3), that the σ_i^l's are positive; that is,

$$\Omega_f = \{\underline{\theta}_f : |\underline{\theta}_f| \leq M_f, \sigma_i^l \geq \sigma\} \tag{8.21}$$

where M_f and σ are positive constants specified by the designer. If we use the fuzzy logic system (8.1), ignore the $\sigma_i^l \geq \sigma$ in (8.21). For Ω_g, in addition to the constraints similar to (8.21), we also require that $\hat{g}(\underline{x}|\underline{\theta}_g)$ must be positive (since $g(\underline{x})$ is positive). Observing (8.1) and (8.3), we have

$$\Omega_g = \{\underline{\theta}_g : |\underline{\theta}_g| \leq M_g, \bar{y}^l \geq \epsilon, \sigma_i^l \geq \sigma\} \tag{8.22}$$

where M_g, ϵ, σ are positive constants specified by the designer. Since both fuzzy logic systems (8.1) and (8.3) are weighted averages of $\bar{y}^l$'s, $\bar{y}^l \geq \epsilon > 0$ implies that the corresponding fuzzy logic systems are positive. If we use the fuzzy logic system (8.1), ignore the $\sigma_i^l \geq \sigma$ constraint. Define the *minimum approximation error*

$$w = (\hat{f}(\underline{x}|\underline{\theta}_f^*) - f(\underline{x})) + (\hat{g}(\underline{x}|\underline{\theta}_g^*) - g(\underline{x}))u_c \tag{8.23}$$

Then the error equation (8.15) can be rewritten as

$$\dot{\underline{e}} = \Lambda_c \underline{e} - \underline{b}_c g(\underline{x}) u_s + \underline{b}_c[(\hat{f}(\underline{x}|\underline{\theta}_f) - \hat{f}(\underline{x}|\underline{\theta}_f^*)) + (\hat{g}(\underline{x}|\underline{\theta}_g) - \hat{g}(\underline{x}|\underline{\theta}_g^*))u_c + w] \tag{8.24}$$

If we choose $\hat{f}$ and $\hat{g}$ to be the fuzzy logic systems in the form of (8.1), then (8.24) can be rewritten as

$$\dot{\underline{e}} = \Lambda_c \underline{e} - \underline{b}_c g(\underline{x}) u_s + \underline{b}_c w + \underline{b}_c[\underline{\phi}_f^T \underline{\xi}(\underline{x}) + \underline{\phi}_g^T \underline{\xi}(\underline{x}) u_c] \tag{8.25}$$

where $\underline{\phi}_f = \underline{\theta}_f - \underline{\theta}_f^*$, $\underline{\phi}_g = \underline{\theta}_g - \underline{\theta}_g^*$, and $\underline{\xi}(\underline{x})$ is the fuzzy basis function (8.2). Now consider the Lyapunov function candidate

$$V = \frac{1}{2}\underline{e}^T P \underline{e} + \frac{1}{2\gamma_1}\underline{\phi}_f^T \underline{\phi}_f + \frac{1}{2\gamma_2}\underline{\phi}_g^T \underline{\phi}_g \tag{8.26}$$

where γ_1 and γ_2 are positive constants. The time derivative of V along the trajectory of (8.25) is

$$\dot{V} = -\frac{1}{2}\underline{e}^T Q \underline{e} - g(\underline{x})\underline{e}^T P \underline{b}_c u_s + \underline{e}^T P \underline{b}_c w$$
$$+ \frac{1}{\gamma_1}\underline{\phi}_f^T[\dot{\underline{\theta}}_f + \gamma_1 \underline{e}^T P \underline{b}_c \underline{\xi}(\underline{x})] + \frac{1}{\gamma_2}\underline{\phi}_g^T[\dot{\underline{\theta}}_g + \gamma_2 \underline{e}^T P \underline{b}_c \underline{\xi}(\underline{x}) u_c] \tag{8.27}$$

where we used (8.12) and $\dot{\underline{\phi}}_f = \dot{\underline{\theta}}_f$, $\dot{\underline{\phi}}_g = \dot{\underline{\theta}}_g$. From (8.17) and $g(\underline{x}) > 0$ we have that $g(\underline{x})\underline{e}^T P \underline{b}_c u_s \geq 0$. If we choose the adaptive law

$$\dot{\underline{\theta}}_f = -\gamma_1 \underline{e}^T P \underline{b}_c \underline{\xi}(\underline{x}) \tag{8.28}$$

$$\dot{\underline{\theta}}_g = -\gamma_2 \underline{e}^T P \underline{b}_c \underline{\xi}(\underline{x}) u_c \tag{8.29}$$

then from (8.27) we have

$$\dot{V} \leq -\frac{1}{2}\underline{e}^T Q \underline{e} + \underline{e}^T P \underline{b}_c w \tag{8.30}$$

This is the best we can hope to get because the term $\underline{e}^T P \underline{b}_c w$ is of the order of the minimum approximation error. If $w = 0$, that is, the searching spaces for $\hat{f}$ and $\hat{g}$ are so big that the f and g are included in them, then we have $\dot{V} \leq 0$. Because of the Universal Approximation Theorem, we can expect that the w should be small, if not equal to zero, provided that we use sufficiently complex (in terms of number of adjustable parameters) $\hat{f}$ and $\hat{g}$.

If we choose $\hat{f}$ and $\hat{g}$ to be the fuzzy logic systems in the form of (8.3), then in order to use the same strategy as previously, we have to approximate $\hat{f}$ and $\hat{g}$ using Taylor series expansions. Specifically, taking the Taylor series expansions of $\hat{f}(\underline{x}|\underline{\theta}_f^*)$ and $\hat{g}(\underline{x}|\underline{\theta}_g^*)$ around $\underline{\theta}_f$ and $\underline{\theta}_g$, we have

$$\hat{f}(\underline{x}|\underline{\theta}_f) - \hat{f}(\underline{x}|\underline{\theta}_f^*) = \underline{\phi}_f^T \left(\frac{\partial \hat{f}(\underline{x}|\underline{\theta}_f)}{\partial \underline{\theta}_f} \right) + O(|\underline{\phi}_f|^2) \tag{8.31}$$

$$\hat{g}(\underline{x}|\underline{\theta}_g) - \hat{g}(\underline{x}|\underline{\theta}_g^*) = \underline{\phi}_g^T \left(\frac{\partial \hat{g}(\underline{x}|\underline{\theta}_g)}{\partial \underline{\theta}_g} \right) + O(|\underline{\phi}_g|^2) \tag{8.32}$$

where $O(|\underline{\phi}_f|^2)$ and $O(|\underline{\phi}_g|^2)$ are the higher-order terms. Substituting (8.31) and (8.32) into (8.24), we have

$$\dot{\underline{e}} = \Lambda_c \underline{e} - \underline{b}_c g(\underline{x}) u_s + \underline{b}_c v + \underline{b}_c [\underline{\phi}_f^T \left(\frac{\partial \hat{f}(\underline{x}|\underline{\theta}_f)}{\partial \underline{\theta}_f} \right) + \underline{\phi}_g^T \left(\frac{\partial \hat{g}(\underline{x}|\underline{\theta}_g)}{\partial \underline{\theta}_g} \right) u_c] \tag{8.33}$$

where

$$v = w + O(|\underline{\phi}_f|^2) + O(|\underline{\phi}_g^2|) u_c \tag{8.34}$$

The rest of the development is the same as (8.26)–(8.30); we omit the details.

The final problem is how to constrain the $\underline{\theta}_f$ and $\underline{\theta}_g$ within the sets Ω_f (8.21) and Ω_g (8.22), respectively. If we can keep $\underline{\theta}_f \in \Omega_f$ and $\underline{\theta}_g \in \Omega_g$, then u_c (8.8) and u_s (8.17) will be bounded because in this case $\hat{f}$ is bounded, $\hat{g} > 0$, and recall that $\underline{e}$ is bounded due to the supervisory control u_s. Clearly, the adaptive law (8.28) and (8.29) cannot guarantee that $\underline{\theta}_f \in \Omega_f$ and $\underline{\theta}_g \in \Omega_g$. To solve this problem, we use the parameter projection algorithm [Luenberger, 39]: if the parameter vectors $\underline{\theta}_f$ and $\underline{\theta}_g$ are within the constraint sets or on the boundaries of the constraint sets but moving toward the inside of the constraint sets, then use the simple adaptive law (8.28) and (8.29). Otherwise, if the parameter vectors are on the boundaries of the constraint sets but moving toward the outside of the constraint sets, then use the projection algorithm to modifiy the adaptive law (8.28) and (8.29) such that the parameter vectors will remain inside the constraint sets. We will show the details in Sections 8.3 and 8.4. The overall control scheme is shown in Figure 8.1.

A Constructive Lyapunov Synthesis Approach to Indirect Fuzzy Controller Design

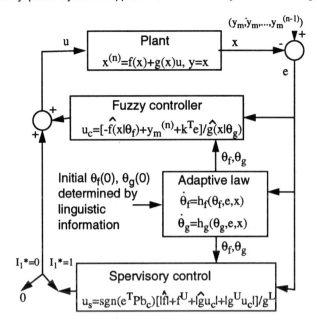

Figure 8.1 The overall scheme of indirect adaptive fuzzy control.

We have shown all the basic ideas of constructing stable indirect adaptive fuzzy controllers in a constructive manner. In Sections 8.3 and 8.4, we will reverse the procedure by first showing the detailed design steps of the adaptive fuzzy controllers, and then proving that the adaptive fuzzy controllers so designed have the desired properties. We think that the way of presentation in this section should make it easier to understand how the adaptive fuzzy controllers are obtained, whereas the way of presentation in Sections 8.3 and 8.4 should make it easier to use them.

Before we conclude this section, we make an assumption.

Assumption 8.2. There are the following linguistic descriptions about the unknown functions $f(\underline{x})$ and $g(\underline{x})$ (from human experts):

$$R_f^{(r)} : IF \ x_1 \ is \ A_1^r \ and \ \cdots \ and \ x_n \ is \ A_n^r,$$
$$THEN \ f(\underline{x}) \ is \ C^r \tag{8.35}$$

and

$$R_g^{(s)} : IF \ x_1 \ is \ B_1^s \ and \ \cdots \ and \ x_n \ is \ B_n^s,$$
$$THEN \ g(\underline{x}) \ is \ D^s \tag{8.36}$$

respectively, where A_i^r, B_i^s, C^r, and D^s are fuzzy sets in R, $r = 1, 2, \ldots, L_f$ and $s = 1, 2, \ldots, L_g$.

We allow $L_f = L_g = 0$, which means that there are no linguistic descriptions (8.35) and (8.36) about $f(\underline{x})$ and $g(\underline{x})$; therefore, Assumption 8.2 is not necessary. We make this assumption for the purpose of emphasizing that our indirect adaptive fuzzy controllers (in Sections 8.3 and 8.4) can directly incorporate these linguistic descriptions (if there are any) into their designs.

8.3 DESIGN AND STABILITY ANALYSIS OF FIRST-TYPE INDIRECT ADAPTIVE FUZZY CONTROLLERS

In this section, we choose $\hat{f}(\underline{x}|\underline{\theta}_f)$ and $\hat{g}(\underline{x}|\underline{\theta}_g)$ to be the fuzzy logic system in the form of (8.1). We first present the detailed design steps of the adaptive fuzzy controller, and then study its properties.

Design of First-Type Indirect Adaptive Fuzzy Controllers

Step 1: Off-line Preprocessing

- Specify the $k_1, \ldots, k_n$ such that all roots of $s^n + k_1 s^{n-1} + \cdots + k_n = 0$ are in the open left-half plane. Specify a positive definite $n \times n$ matrix Q.
- Solve the Lyapunov equation (8.12), for example, using the method in Wang and Mendel [90], to obtain a symmetric $P > 0$.
- Specify the design parameters M_f, M_g, ϵ, and $\bar{V}$ based on practical constraints (see Remark 8.1 for further discussion).

Step 2: Initial Controller Construction

- Define m_i fuzzy sets $F_i^{l_i}$ whose membership functions $\mu_{F_i^{l_i}}$ uniformly cover U_{c_i} which is the projection of U_c onto the ith coordinate, where $l_i = 1, 2, \ldots, m_i$ and $i = 1, 2, \ldots, n$. We require that the $F_i^{l_i}$'s include the A_i^r's and B_i^s's in (8.35) and (8.36).
- Construct the fuzzy rule bases for the fuzzy logic systems $\hat{f}(\underline{x}|\underline{\theta}_f)$ and $\hat{g}(\underline{x}|\underline{\theta}_g)$, each of which consists of $m_1 \times m_2 \times \cdots \times m_n$ rules whose IF parts comprise all the possible combinations of the $F_i^{l_i}$'s for $i = 1, 2, \ldots, n$. Specifically, the fuzzy rule bases of $\hat{f}(\underline{x}|\underline{\theta}_f)$ and $\hat{g}(\underline{x}|\underline{\theta}_g)$ consist of rules

$$R_f^{(l_1,\ldots,l_n)} : \quad IF \quad x_1 \text{ is } F_1^{l_1} \text{ and } \cdots \text{ and } x_n \text{ is } F_n^{l_n},$$

$$THEN \quad \hat{f}(\underline{x}|\underline{\theta}_f) \text{ is } G^{(l_1,\ldots,l_n)} \qquad (8.37)$$

$$R_g^{(l_1,\ldots,l_n)} : \quad IF \quad x_1 \text{ is } F_1^{l_1} \text{ and } \cdots \text{ and } x_n \text{ is } F_n^{l_n},$$

$$THEN \quad \hat{g}(\underline{x}|\underline{\theta}_g) \text{ is } H^{(l_1,\ldots,l_n)} \qquad (8.38)$$

Design and Stability Analysis of First-Type Indirect Adaptive Fuzzy Controllers 115

respectively, where $l_i = 1, 2, \ldots, m_i$, $i = 1, 2, \ldots, n$, and $G^{(l_1,\ldots,l_n)}$ and $H^{(l_1,\ldots,l_n)}$ are fuzzy sets in R which are specified as follows: if the IF part of (8.37) or (8.38) agrees with the IF part of (8.35) or (8.36), set $G^{(l_1,\ldots,l_n)}$ or $H^{(l_1,\ldots,l_n)}$ equal to the corresponding C^r or D^s, respectively; otherwise, set $G^{(l_1,\ldots,l_n)}$ and $H^{(l_1,\ldots,l_n)}$ arbitrarily with the constraint that the centers of $G^{(l_1,\ldots,l_n)}$ and $H^{(l_1,\ldots,l_n)}$ (which correspond to the $\bar{y}^l$ parameters) are inside the constraint sets Ω_f and Ω_g, respectively. Therefore, *the initial adaptive fuzzy controller is constructed from the linguistic rules (8.35) and (8.36)*.

- Construct the fuzzy basis functions

$$\xi^{(l_1,\ldots,l_n)}(\underline{x}) = \frac{\prod_{i=1}^{n} \mu_{F_i^{l_i}}(x_i)}{\sum_{l_1=1}^{m_1} \cdots \sum_{l_n=1}^{m_n} (\prod_{i=1}^{n} \mu_{F_i^{l_i}}(x_i))} \tag{8.39}$$

and collect them into a $\prod_{i=1}^{n} m_i$-dimensional vector $\underline{\xi}(\underline{x})$ in a natural ordering for $l_1 = 1, 2, \ldots, m_1, \ldots, l_n = 1, 2, \ldots, m_n$. Collect the points at which $\mu_{G^{(l_1,\ldots,l_n)}}$ and $\mu_{H^{(l_1,\ldots,l_n)}}$ achieve their maximum values, in the same ordering as $\underline{\xi}(\underline{x})$, into vectors $\underline{\theta}_f(0)$ and $\underline{\theta}_g(0)$, respectively. The $\hat{f}(\underline{x}|\underline{\theta}_f)$ and $\hat{g}(\underline{x}|\underline{\theta}_g)$ are constructed as

$$\hat{f}(\underline{x}|\underline{\theta}_f) = \underline{\theta}_f^T \underline{\xi}(\underline{x}) \tag{8.40}$$

$$\hat{g}(\underline{x}|\underline{\theta}_g) = \underline{\theta}_g^T \underline{\xi}(\underline{x}) \tag{8.41}$$

Step 3: On-line Adaptation

- Apply the feedback control (8.14) to the plant (8.5), where u_c is given by (8.8), u_s is given by (8.17), and $\hat{f}(\underline{x}|\underline{\theta}_f)$ and $\hat{g}(\underline{x}|\underline{\theta}_g)$ are given by (8.40) and (8.41), respectively.
- Use the following adaptive law to adjust the parameter vector $\underline{\theta}_f$:

$$\dot{\underline{\theta}}_f = \begin{cases} -\gamma_1 \underline{e}^T P \underline{b}_c \underline{\xi}(\underline{x}) & \text{if } (|\underline{\theta}_f| < M_f) \text{ or } (|\underline{\theta}_f| = M_f \text{ and } \underline{e}^T P \underline{b}_c \underline{\theta}_f^T \underline{\xi}(\underline{x}) \geq 0) \\ P\{-\gamma_1 \underline{e}^T P \underline{b}_c \underline{\xi}(\underline{x})\} & \text{if } (|\underline{\theta}_f| = M_f \text{ and } \underline{e}^T P \underline{b}_c \underline{\theta}_f^T \underline{\xi}(\underline{x}) < 0) \end{cases} \tag{8.42}$$

where the projection operator $P\{*\}$ is defined as [Luenberger, 39]

$$P\{-\gamma_1 \underline{e}^T P \underline{b}_c \underline{\xi}(\underline{x})\} = -\gamma_1 \underline{e}^T P \underline{b}_c \underline{\xi}(\underline{x}) + \gamma_1 \underline{e}^T P \underline{b}_c \frac{\underline{\theta}_f \underline{\theta}_f^T \underline{\xi}(\underline{x})}{|\underline{\theta}_f|^2} \tag{8.43}$$

- Use the following adaptive law to adjust the parameter vector $\underline{\theta}_g$:
  □ Whenever an element θ_{gi} of $\underline{\theta}_g = \epsilon$, use

$$\dot{\theta}_{gi} = \begin{cases} -\gamma_2 \underline{e}^T P \underline{b}_c \xi_i(\underline{x}) u_c & \text{if } \underline{e}^T P \underline{b}_c \xi_i(\underline{x}) u_c < 0 \\ 0 & \text{if } \underline{e}^T P \underline{b}_c \xi_i(\underline{x}) u_c \geq 0 \end{cases} \tag{8.44}$$

where $\xi_i(\underline{x})$ is the ith component of $\underline{\xi}(\underline{x})$.

☐ Otherwise, use

$$\dot{\underline{\theta}}_g = \begin{array}{l} -\gamma_2 \underline{e}^T P \underline{b}_c \underline{\xi}(\underline{x}) u_c \text{ if } (|\underline{\theta}_g| < M_g) \text{ or } (|\underline{\theta}_g| = M_g \text{ and } \underline{e}^T P \underline{b}_c \underline{\theta}_g^T \underline{\xi}(\underline{x}) u_c \geq 0) \\ \\ P\{-\gamma_2 \underline{e}^T P \underline{b}_c \underline{\xi}(\underline{x}) u_c\} \text{ if } (|\underline{\theta}_g| = M_g \text{ and } \underline{e}^T P \underline{b}_c \underline{\theta}_g^T \underline{\xi}(\underline{x}) u_c < 0) \end{array} \quad (8.45)$$

where the projection operator $P\{*\}$ is defined as

$$P\{-\gamma_2 \underline{e}^T P \underline{b}_c \underline{\xi}(\underline{x}) u_c\} = -\gamma_2 \underline{e}^T P \underline{b}_c \underline{\xi}(\underline{x}) u_c + \gamma_2 \underline{e}^T P \underline{b}_c \frac{\underline{\theta}_g \underline{\theta}_g^T \underline{\xi}(\underline{x}) u_c}{|\underline{\theta}_g|^2} \quad (8.46)$$

The following theorem shows the properties of this adaptive fuzzy controller.

Theorem 8.1. Consider the plant (8.5) with the control (8.14), where u_c is given by (8.8), u_s is given by (8.17), and $\hat{f}$ and $\hat{g}$ are given by (8.40) and (8.41), respectively. Let the parameter vectors $\underline{\theta}_f$ and $\underline{\theta}_g$ be adjusted by the adaptive law (8.42)–(8.46), and let Assumptions 8.1 and 8.2 be true. Then, the overall control scheme (shown in Figure 8.1) guarantees the following properties:

1. $|\underline{\theta}_f(t)| \leq M_f$, $|\underline{\theta}_g(t)| \leq M_g$, all elements of $\underline{\theta}_g \geq \epsilon$,

$$|\underline{x}(t)| \leq |\underline{y}_m| + (\frac{2\bar{V}}{\lambda_{min}})^{1/2} \quad (8.47)$$

and

$$|u(t)| \leq \frac{1}{\epsilon}(M_f + |y_m^{(n)}| + |\underline{k}|(\frac{2\bar{V}}{\lambda_{min}})^{1/2})$$
$$+ \frac{1}{g_L(\underline{x})}[M_f + |f^U(\underline{x})|$$
$$+ \frac{1}{\epsilon}(M_g + g^U)(M_f + |y_m^{(n)}| + |\underline{k}|(\frac{2\bar{V}}{\lambda_{min}})^{1/2})] \quad (8.48)$$

for all $t \geq 0$, where λ_{min} is the minimum eigenvalue of P, and $\underline{y}_m = (y_m, \dot{y}_m, \ldots, y_m^{(n-1)})^T$.

2.
$$\int_0^t |\underline{e}(\tau)|^2 d\tau \leq a + b \int_0^t |w(\tau)|^2 d\tau \quad (8.49)$$

for all $t \geq 0$, where a and b are constants, and w is minimum approximation error defined by (8.23).

3. If w is squared integrable, that is, $\int_0^\infty |w(t)|^2 dt < \infty$, then $lim_{t \to \infty} |\underline{e}(t)| = 0$.

Proof of this theorem is given in the Appendix. We now make a few remarks on this adaptive fuzzy controller.

Remark 8.1. For many practical control problems, the state $\underline{x}$ and control u are required to be constrained within certain regions. For given constraints, we can specify the design parameters $\underline{k}$, M_f, M_g, ϵ, and $\bar{V}$, based on (8.47) and (8.48), such that the state $\underline{x}$ and control u are within the constraint sets. To do this, we need to know some fixed bounds of $|\underline{y}_m|$, $|y_m^{(n)}|$, $|f^U(\underline{x})|$, $g^U(\underline{x})$, and $g_L(\underline{x})$. Since these functions are known to the designer, it should not be difficult to determine these bounds. After these bounds are determined, we can specify the values of the right-hand sides of (8.47) and (8.48) by properly choosing the design parameters. Note from (8.12) and (8.11) that λ_{min} is determined by $\underline{k}$. Therefore, we can specify $\underline{k}$ to achieve a required λ_{min}. In Section 8.5, we will show an example (the inverted pendulum tracking control) of how to specify the design parameters such that the state and control are within given constraint sets.

Remark 8.2. If we choose $I_1^* \equiv 1$ in the u_s of (8.17), then from (8.18) we guarantee to have not only the boundedness of the state vector but also the error e converge to zero. We do not choose this strategy because the u_s is usually very large. Indeed, from (8.17) we see that the u_s is propositional to the upper bounds f^U and g^U which are usually very large. Large control is undesirable because it may increase the implementation cost. We therefore choose the u_s to operate in the preceding supervisory fashion.

Remark 8.3. From Step 2 we see that the linguistic information (8.35) and (8.36) is incorporated into the adaptive fuzzy controller by constructing the initial controller based on it. If the linguistic rules (8.35) and (8.36) provide good descriptions about $f(\underline{x})$ and $g(\underline{x})$, then the initial $\hat{f}$ and $\hat{g}$ should be close to the true $f(\underline{x})$ and $g(\underline{x})$, respectively; as a result, we can hope that the closed-loop system behaves approximately like (8.7). If no linguistic information is available, our adaptive fuzzy controller is still a well-performing nonlinear adaptive controller, in the sense of having the listed properties of Theorem 8.1. In summary, *good linguistic information can help us to construct a good initial controller so that we can have a fast adaptation*; we will show an example in Section 8.5 to illustrate this point.

Remark 8.4. From the third item of Theorem 8.1 we see that in order for the tracking error $e(t)$ to converge to zero, we require that the "minimum approximation error w" defined by (8.23) is small (in the sense of squared integrable). From the Universal Approximation Theorem we can say that if we use a sufficient number of rules to construct $\hat{f}$ and $\hat{g}$, the w should be small.

Remark 8.5. The basic idea of the projection algorithm in (8.42)–(8.46) is as follows: if the parameter vector is inside the constraint set or on the boundary of the constraint set but moving toward the inside of the constraint set (which corresponds to the cases of the first lines of (8.42), (8.44), and (8.45)), then use the simple adaptive law based on the Lyapunov synthesis approach (see (8.28) and (8.29)); if the parameter vector is on the boundary of the constraint set but moving

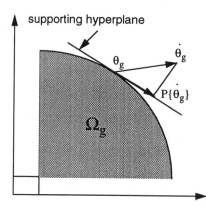

Figure 8.2 Illustration of the projection algorithm.

toward the outside of the constraint set (which corresponds to the cases of the second lines of (8.42), (8.44), and (8.45)), then project the gradient vector $\dot{\underline{\theta}}_f$ or $\dot{\underline{\theta}}_g$ onto the supporting hyperplane [Luenberger, 39] at $\underline{\theta}_f$ or $\underline{\theta}_g$ to the convex set Ω_f or Ω_g. Figure 8.2 shows a two-dimensional example for $\underline{\theta}_g$.

Remark 8.6. In this adaptive fuzzy controller, we fix the fuzzy sets in the IF parts of the rules for the fuzzy systems $\hat{f}$ and $\hat{g}$. An advantage of doing so is that the fuzzy systems $\hat{f}$ and $\hat{g}$ are linear in the parameter. Therefore (1) we were able to use a relatively simpler adaptive law to adjust the parameters, and (2) the convergence of the adaptation procedure is expected to be faster because we are not concerned with complicated nonlinear search problems. A disadvantage is that we have to consider all the possible combinations of the fuzzy sets in U_c because these fuzzy sets cannot change so that we should have rules to cover every region of U_c, where by "cover" we mean that for each $\underline{x} \in U_c$ there should be at least one rule in the fuzzy rule bases of $\hat{f}$ and $\hat{g}$ whose "strength" $(\mu_{F_1^{l_1}}(x_1) \cdots \mu_{F_n^{l_n}}(x_n))$ is not very small. Since in general the real trajectory of $\underline{x}(t)$ is only in a certain small region of U_c, many rules in $\hat{f}$ and $\hat{g}$ are not used for an implementation of the adaptive fuzzy controller; that is, this adaptive fuzzy controller is not efficient in utilizing its adjustable parameters. To overcome this disadvantage, we develop another adaptive fuzzy controller next in which the fuzzy sets in the IF parts are also adjustable during the adaptation procedure.

8.4 DESIGN AND STABILITY ANALYSIS OF SECOND-TYPE INDIRECT ADAPTIVE FUZZY CONTROLLERS

From Remark 8.6 we see that the first-type indirect adaptive fuzzy controller may require a large number of rules for higher-dimensional systems. A way to overcome this rule explosion problem is to allow the fuzzy sets in the IF parts of the rules also to change during the adaptation procedure so that in principle any rule can cover any region of U_c; as a result, we only need a small number of rules. Specifically, we will choose $\hat{f}$ and $\hat{g}$ to be the fuzzy systems in the form of (8.3), and develop an adaptive law to adjust all the parameters $\bar{y}^l, \bar{x}_i^l$, and σ_i^l. The price paid for this additional freedom is that the fuzzy systems $\hat{f}$ and $\hat{g}$ become nonlinear in the parameter, so that we have to use a more complicated adaptive law.

Design of Second-Type Indirect Adaptive Fuzzy Controllers

Step 1: Off-line Preprocessing

This is the same as the adaptive fuzzy controller in the previous section, except that we need to specify one more design parameter σ.

Step 2: Initial Controller Construction

- Choose $\hat{f}(\underline{x}|\underline{\theta}_f)$ and $\hat{g}(\underline{x}|\underline{\theta}_g)$ to be the fuzzy logic systems in the form of (8.3); that is,

$$\hat{f}(\underline{x}|\underline{\theta}_f) = \frac{\sum_{l=1}^{M} \bar{y}_f^l \left[\prod_{i=1}^{n} exp\left(-\left(\frac{x_i - \bar{x}_{fi}^l}{\sigma_{fi}^l} \right)^2 \right) \right]}{\sum_{l=1}^{M} \left[\prod_{i=1}^{n} exp\left(-\left(\frac{x_i - \bar{x}_{fi}^l}{\sigma_{fi}^l} \right)^2 \right) \right]} \quad (8.50)$$

$$\hat{g}(\underline{x}|\underline{\theta}_g) = \frac{\sum_{l=1}^{M} \bar{y}_g^l \left[\prod_{i=1}^{n} exp\left(-\left(\frac{x_i - \bar{x}_{gi}^l}{\sigma_{gi}^l} \right)^2 \right) \right]}{\sum_{l=1}^{M} \left[\prod_{i=1}^{n} exp\left(-\left(\frac{x_i - \bar{x}_{gi}^l}{\sigma_{gi}^l} \right)^2 \right) \right]} \quad (8.51)$$

where $\underline{\theta}_f$ ($\underline{\theta}_g$) is the collection of the adjustable parameters $\bar{y}_f^l, \bar{x}_{fi}^l$, and σ_{fi}^l ($\bar{y}_g^l, \bar{x}_{gi}^l$ and σ_{gi}^l). Clearly, $\hat{f}$ and $\hat{g}$ in (8.50) and (8.51) are constructed based on M rules whose IF parts are characterized by the Gaussian membership functions

$$\mu_{F_i^l}(x_i) = exp\left(-\left(\frac{x_i - \bar{x}_{fi}^l}{\sigma_{fi}^l} \right)^2 \right) \quad (8.52)$$

$$\mu_{G_i^l}(x_i) = exp\left(-\left(\frac{x_i - \bar{x}_{gi}^l}{\sigma_{gi}^l} \right)^2 \right) \quad (8.53)$$

respectively, where $l = 1, 2, \ldots, M$ and $i = 1, 2, \ldots, n$. For this adaptive fuzzy controller, we assume that the A_i^r and B_i^s in (8.35) and (8.36) are also characterized by the Gaussian membership functions in the form of (8.52) and (8.53), and that $L_f \leq M$ and $L_g \leq M$.

- Determine the initial $\underline{\theta}_f(0)$ and $\underline{\theta}_g(0)$ as follows: for the L_f (L_g) $\mu_{F_i^l}$'s ($\mu_{G_i^l}$'s) that are the same as the $\mu_{A_i^r}$'s ($\mu_{B_i^s}$'s) in (8.35) ((8.36)), determine the $\bar{x}_{fi}^l(0)$ and $\sigma_{fi}^l(0)$ ($\bar{x}_{gi}^l(0)$ and $\sigma_{gi}^l(0)$) based on the $\mu_{A_i^r}$'s ($\mu_{B_i^s}$'s), and choose the $\bar{y}_f^l(0)$'s ($\bar{y}_g^l(0)$'s) to be the centers of the corresponding μ_{C^r}'s (μ_{D^s}'s); the remaining parameters are chosen arbitrarily in the constraint sets (8.21) and (8.22).

Step 3: On-line Adaptation

- Compute $\frac{\partial \hat{f}(\underline{x}|\underline{\theta}_f)}{\partial \underline{\theta}_f}$ using the following algorithm:

$$\frac{\partial \hat{f}}{\partial \bar{y}_f^l} = \frac{b_f^l}{\sum_{l=1}^{M} b_f^l} \tag{8.54}$$

$$\frac{\partial \hat{f}}{\partial \bar{x}_{fi}^l} = \frac{\bar{y}_f^l - \hat{f}}{\sum_{l=1}^{M} b_f^l} b_f^l \frac{-2(x_i - \bar{x}_{fi}^l)}{(\sigma_{fi}^l)^2} \tag{8.55}$$

$$\frac{\partial \hat{f}}{\partial \sigma_{fi}^l} = \frac{\bar{y}_f^l - \hat{f}}{\sum_{l=1}^{M} b_f^l} b_f^l \frac{-2(x_i - \bar{x}_{fi}^l)^2}{(\sigma_{fi}^l)^3} \tag{8.56}$$

where

$$b_f^l = \prod_{i=1}^{n} exp\left(-\left(\frac{x_i - \bar{x}_{fi}^l}{\sigma_{fi}^l}\right)^2\right) \tag{8.57}$$

(8.54)–(8.56) are obtained by taking differentials of $\hat{f}$ in (8.50) with respect to the corresponding parameters (see Chapter 3). Compute $\frac{\partial \hat{g}(\underline{x}|\underline{\theta}_g)}{\partial \underline{\theta}_g}$ using the same algorithm (8.54)–(8.56), replacing f by g.

- Apply the feedback control (8.14) to the plant (8.5), where u_c is given by (8.8), u_s is given by (8.17), and $\hat{f}(\underline{x}|\underline{\theta}_f)$ and $\hat{g}(\underline{x}|\underline{\theta}_g)$ are given by (8.50) and (8.51), respectively.
- Use the following adaptive law to adjust the parameter vector $\underline{\theta}_f$:
  □ Whenever $\sigma_{fi}^l = \sigma$, use

$$\dot{\sigma}_{fi}^l = \begin{cases} -\gamma_1 \underline{e}^T P \underline{b}_c \frac{\partial \hat{f}}{\partial \sigma_{fi}^l} & if \ \underline{e}^T P \underline{b}_c \frac{\partial \hat{f}}{\partial \sigma_{fi}^l} < 0 \\ \\ 0 & if \ \underline{e}^T P \underline{b}_c \frac{\partial \hat{f}}{\partial \sigma_{fi}^l} \geq 0 \end{cases} \tag{8.58}$$

□ Otherwise, use

$$\dot{\underline{\theta}}_f = \begin{cases} -\gamma_1 \underline{e}^T P \underline{b}_c \frac{\partial \hat{f}}{\partial \underline{\theta}_f} & if \ (|\underline{\theta}_f| < M_f) \\ & or \ (|\underline{\theta}_f| = M_f \ and \ \underline{e}^T P \underline{b}_c \underline{\theta}_f^T \frac{\partial \hat{f}}{\partial \underline{\theta}_f} \geq 0) \\ \\ -\gamma_1 \underline{e}^T P \underline{b}_c \frac{\partial \hat{f}}{\partial \underline{\theta}_f} + \gamma_1 \underline{e}^T P \underline{b}_c \frac{\underline{\theta}_f \underline{\theta}_f^T}{|\underline{\theta}_f|^2} \frac{\partial \hat{f}}{\partial \underline{\theta}_f} \\ & if \ (|\underline{\theta}_f| = M_f \ and \ \underline{e}^T P \underline{b}_c \underline{\theta}_f^T \frac{\partial \hat{f}}{\partial \underline{\theta}_f} < 0) \end{cases} \tag{8.59}$$

Design and Stability Analysis of Second-Type Indirect Adaptive Fuzzy Controllers

- Use the following adaptive law to adjust the parameter vector $\underline{\theta}_g$:
  - Whenever $\bar{y}_g^l = \epsilon$, use

$$\dot{\bar{y}}_g^l = \begin{cases} -\gamma_2 \underline{e}^T P \underline{b}_c \dfrac{\partial \hat{g}}{\partial \bar{y}_g^l} u_c & \text{if } \underline{e}^T P \underline{b}_c \dfrac{\partial \hat{g}}{\partial \bar{y}_g^l} u_c < 0 \\ \\ 0 & \text{if } \underline{e}^T P \underline{b}_c \dfrac{\partial \hat{g}}{\partial \bar{y}_g^l} u_c \geq 0 \end{cases} \quad (8.60)$$

  - Whenever $\sigma_{gi}^l = \sigma$, use

$$\dot{\sigma}_{gi}^l = \begin{cases} -\gamma_2 \underline{e}^T P \underline{b}_c \dfrac{\partial \hat{g}}{\partial \sigma_{gi}^l} u_c & \text{if } \underline{e}^T P \underline{b}_c \dfrac{\partial \hat{g}}{\partial \sigma_{gi}^l} u_c < 0 \\ \\ 0 & \text{if } \underline{e}^T P \underline{b}_c \dfrac{\partial \hat{g}}{\partial \sigma_{gi}^l} u_c \geq 0 \end{cases} \quad (8.61)$$

  - Otherwise, use

$$\dot{\underline{\theta}}_g = \begin{cases} -\gamma_2 \underline{e}^T P \underline{b}_c \dfrac{\partial \hat{g}}{\partial \underline{\theta}_g} u_c & \text{if } (|\underline{\theta}_g| < M_g) \\ & \text{or } (|\underline{\theta}_g| = M_g \text{ and } \underline{e}^T P \underline{b}_c \underline{\theta}_g^T \dfrac{\partial \hat{g}}{\partial \underline{\theta}_g} u_c \geq 0) \\ \\ -\gamma_2 \underline{e}^T P \underline{b}_c \dfrac{\partial \hat{g}}{\partial \underline{\theta}_g} u_c + \gamma_2 \underline{e}^T P \underline{b}_c \dfrac{\underline{\theta}_g \underline{\theta}_g^T}{|\underline{\theta}_g|^2} \dfrac{\partial \hat{g}}{\partial \underline{\theta}_g} u_c \\ & \text{if } (|\underline{\theta}_g| = M_g \text{ and } \underline{e}^T P \underline{b}_c \underline{\theta}_g^T \dfrac{\partial \hat{g}}{\partial \underline{\theta}_g} u_c < 0) \end{cases} \quad (8.62)$$

Properties of this second-type indirect adaptive fuzzy controller are summarized in the following theorem.

Theorem 8.2. The second-type indirect adaptive fuzzy controller designed through the preceding three steps guarantees the following properties:

1. $|\underline{\theta}_f| \leq M_f$, the σ_{fi}^l's in $\underline{\theta}_f \geq \sigma$, $|\underline{\theta}_g| \leq M_g$, the σ_{gi}^l's in $\underline{\theta}_g \geq \sigma$, the $\bar{y}_g^l$'s in $\underline{\theta}_g \geq \epsilon$, and $\underline{x}$ and u satisfy (8.47) and (8.48), respectively.
2.
$$\int_0^t |\underline{e}(\tau)|^2 d\tau \leq a + b \int_0^t |v(\tau)|^2 d\tau \quad (8.63)$$

for all $t \geq 0$, where a and b are constants, and v is defined by (8.34).

3. If v is squared integrable, then $lim_{t\to\infty}|\underline{e}(t)| = 0$.

Proof of this theorem is given in the Appendix.

Remark 8.7. Remarks 8.1–8.5 apply to this second-type indirect adaptive fuzzy controller.

Remark 8.8. From (8.34) we see that in addition to the minimum approximation error w we have another error term $O(|\underline{\theta}_f - \underline{\theta}_f^*|^2) + O(|\underline{\theta}_g - \underline{\theta}_g^*|^2)u_c$ in this second-type indirect adaptive fuzzy controller. Because in this controller we have a larger searching space for $\hat{f}(\underline{x}|\underline{\theta}_f^*)$ and $\hat{g}(\underline{x}|\underline{\theta}_g^*)$ than in the first-type controller, the minimum approximation error w here should be smaller than the w in the first-type controller. Therefore, the performance of this controller should be more sensitive to the initial $\underline{\theta}_f(0)$ and $\underline{\theta}_g(0)$ than that of the first-type controller. That is, if the initial $\underline{\theta}_f(0)$ and $\underline{\theta}_g(0)$ are close to the optimal $\underline{\theta}_f^*$ and $\underline{\theta}_g^*$, respectively, then the total error v may be smaller than the w in the first-type controller. On the other hand, if the initial $\underline{\theta}_f(0)$ and $\underline{\theta}_g(0)$ are far away from the optimal values, the v will be large. Because the initial controllers are constructed from the linguistic rules (8.35) and (8.36), these rules are more important for the second-type indirect adaptive fuzzy controller than for the first-type indirect adaptive fuzzy controller.

8.5 APPLICATION TO INVERTED PENDULUM TRACKING CONTROL

In this section, we use our two indirect adaptive fuzzy controllers to control the inverted pendulum to track a sine-wave trajectory. Figure 8.3 shows the inverted pendulum system (or the cart-pole system). Let $x_1 = \theta$ and $x_2 = \dot{\theta}$, the dynamic equations of the inverted pendulum system are [Slotine and Li, 66]

$$\dot{x}_1 = x_2$$

$$\dot{x}_2 = \frac{gsinx_1 - \frac{mlx_2^2 cosx_1 sinx_1}{m_c+m}}{l(\frac{4}{3} - \frac{mcos^2 x_1}{m_c+m})} + \frac{\frac{cosx_1}{m_c+m}}{l(\frac{4}{3} - \frac{mcos^2 x_1}{m_c+m})}u \quad (8.64)$$

where $g = 9.8m/s^2$ is the acceleration due to gravity, m_c is the mass of the cart, m is the mass of the pole, l is the half-length of the pole, and u is the applied force (control). We chose $m_c = 1kg$, $m = 0.1kg$, and $l = 0.5m$ in the following simulations. Clearly, (8.64) is in the form of (8.4); thus, our adaptive fuzzy controllers apply to this system. We chose the reference signal $y_m(t) = \frac{\pi}{30}sin(t)$ in the following simulations (other choices are possible).

To apply the adaptive fuzzy controllers to this system, we first need to determine the bounds f^U, g^U, and g_L. For this system, we have

Sec. 8.5 Application to Inverted Pendulum Tracking Control

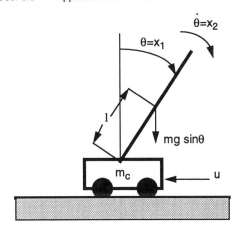

Figure 8.3 The inverted pendulum system.

$$|f(x_1, x_2)| = \left| \frac{gsinx_1 - \frac{mlx_2^2 cosx_1 sinx_1}{m_c+m}}{l\left(\frac{4}{3} - \frac{mcos^2x_1}{m_c+m}\right)} \right|$$

$$\leq \frac{9.8 + \frac{0.025}{1.1}x_2^2}{\frac{2}{3} - \frac{0.05}{1.1}} = 15.78 + 0.0366x_2^2 \equiv f^U(x_1, x_2) \quad (8.65)$$

$$|g(x_1, x_2)| = \left| \frac{cosx_1 \frac{1}{m_c+m}}{l\left(\frac{4}{3} - \frac{mcos^2x_1}{m_c+m}\right)} \right|$$

$$\leq \frac{1}{1.1(\frac{2}{3} - \frac{0.05}{1.1})} = 1.46 \equiv g^U(x_1, x_2) \quad (8.66)$$

If we require that $|x_1| \leq \pi/6$ (we will specify the design parameters such that this requirement is satisfied), then

$$|g(x_1, x_2)| \geq \frac{cos\pi/6}{1.1(\frac{2}{3} + \frac{0.05}{1.1}cos^2\pi/6)} = 1.12 \equiv g_L(x_1, x_2) \quad (8.67)$$

Now suppose we require that

$$|x_1| \leq \pi/6 \quad and \quad |u| \leq 180 \quad (8.68)$$

Since $|x_1| \leq (|x_1|^2 + |x_2|^2)^{1/2} = |\underline{x}|$, if we can make $|\underline{x}| \leq \pi/6$, then $|x_1| \leq \pi/6$. In this case we also have $|x_2| \leq \pi/6$. Our first task is to determine the design parameters $\bar{V}, k_1, k_2, \epsilon, M_f$, and M_g, according to (8.47) and (8.48), such that the constraint (8.68) is satisfied. Since $|y_m| \leq \pi/30$, if we determine $\bar{V}$ and λ_{min} such that $(\frac{2\bar{V}}{\lambda_{min}})^{1/2} \leq 2\pi/15$, then according to (8.47) we have $|\underline{x}| \leq \pi/30 + 2\pi/15 = \pi/6$. Since the number of design parameters is larger than the number of constraints, we have freedom in choosing the design parameters. We simply choose $k_1 = 2$ and $k_2 = 1$ (so that $s^2 + k_1 s + k_2$ is stable) and $Q = diag(10, 10)$. Then, we solve (8.12) and obtain

$$P = \begin{bmatrix} 15 & 5 \\ 5 & 5 \end{bmatrix} \qquad (8.69)$$

This P is positive definite with $\lambda_{min} = 2.93$. To satisfy the constraint for $|\underline{x}|$, we choose $\bar{V} = \frac{\lambda_{min}}{2}(\frac{2\pi}{15})^2 = 0.267$. Finally, we determine M_f and ϵ such that $|u| \leq 180$, according to (8.48). Again, we have additional freedom in choosing the M_f and ϵ. After trial and error, we choose $M_f = 16$, $M_g = 1.6$, and $\epsilon = 0.7$. It is straightforward to verify from (8.47) and (8.48) that the preceding choice of design parameters guarantees that the state and control satisfy (8.68). For the second-type indirect adaptive fuzzy controller, we choose $\sigma = 10^{-2}$.

Now we have finished the off-line preprocessing, that is, Step 1 of the designs of both indirect adaptive fuzzy controllers. Next we simulate the two indirect adaptive fuzzy controllers for this inverted pendulum tracking control problem, each for two cases: without any linguistic rules (Examples 8.1 and 8.3), and with some linguistic rules (Examples 8.2 and 8.4).

Example 8.1. In this example we used the first-type indirect adaptive fuzzy controller, assuming that there are no linguistic rules (8.35) and (8.36). We chose $m_1 = m_2 = 5$. Since $|x_i| \leq \pi/6$ for both $i = 1, 2$, we chose $\mu_{F_i^1}(x_i) = exp[-(\frac{x_i+\pi/6}{\pi/24})^2]$, $\mu_{F_i^2}(x_i) = exp[-(\frac{x_i+\pi/12}{\pi/24})^2]$, $\mu_{F_i^3}(x_i) = exp[-(\frac{x_i}{\pi/24})^2]$, $\mu_{F_i^4}(x_i) = exp[-(\frac{x_i-\pi/12}{\pi/24})^2]$, and $\mu_{F_i^5}(x_i) = exp[-(\frac{x_i-\pi/6}{\pi/24})^2]$, which clearly cover the interval $[-\pi/6, \pi/6]$. From the bounds (8.65)–(8.67) of $f(x_1, x_2)$ and $g(x_1, x_2)$ we see that the range of $f(x_1, x_2)$ is much larger than that of $g(x_1, x_2)$; therefore, we chose $\gamma_1 = 50$ and $\gamma_2 = 1$. We use the same γ_1 and γ_2 in Examples 8.2–8.4. Figures 8.4–8.6 show the simulation results for the initial condition $\underline{x}(0) = (-\frac{\pi}{60}, 0)^T$, where Figure 8.4 shows the state $x_1(t)$ (solid line) and its desired value $y_m(t) = \frac{\pi}{30}sin(t)$ (dashed line), Figure 8.5 shows the state $x_2(t)$ (solid line) and its desired value $\dot{y}_m(t) = \frac{\pi}{30}cos(t)$ (dashed line), and Figure 8.6 shows the control $u(t)$. We chose the initial $\underline{\theta}_f(0)$ randomly in the interval $[-3, 3]$ and $\underline{\theta}_g(0)$ randomly in the interval $[1, 1.3]$. Figures 7.7–8.9 show the same results for the initial condition $\underline{x}(0) = (\frac{\pi}{60}, 0)^T$.

From Figures 8.4, 8.5, 8.7, and 8.8 we see that our first-type indirect adaptive fuzzy controller can control the inverted pendulum to follow the desired trajectory without using any linguistic information. From Figures 8.6 and 8.9 we see that the real control is much less than the bound setting by (8.68); therefore, (8.48) provides a very loose bound. It is certainly of importance to derive a more tight bound than (8.48).

Example 8.2. Here we consider the same situation as in Example 8.1 except that there are some linguistic rules about $f(x_1, x_2)$ and $g(x_1, x_2)$ based on the following physical intuition. First, suppose that there is no control; that is, $u = 0$.

Sec. 8.5 Application to Inverted Pendulum Tracking Control   125

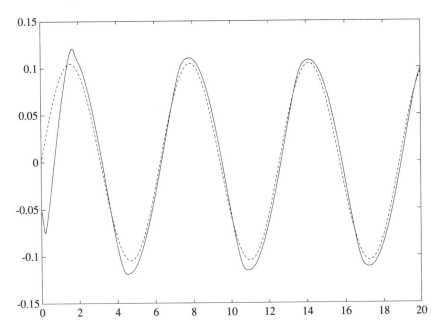

Figure 8.4 The state $x_1(t)$ (solid line) and its desired value $y_m(t) = \frac{\pi}{30}sin(t)$ (dashed line) for the initial condition $\underline{x}(0) = (-\frac{\pi}{60}, 0)^T$ in Example 8.1.

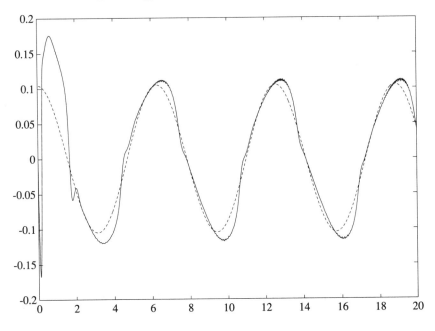

Figure 8.5 The state $x_2(t)$ (solid line) and its desired value $\dot{y}_m(t) = \frac{\pi}{30}cos(t)$ (dashed line) for the initial condition $\underline{x}(0) = (-\frac{\pi}{60}, 0)^T$ in Example 8.1.

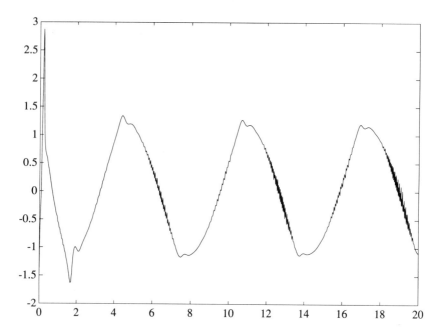

Figure 8.6 The control $u(t)$ for the initial condition $\underline{x}(0) = (-\frac{\pi}{60}, 0)^T$ in Example 8.1.

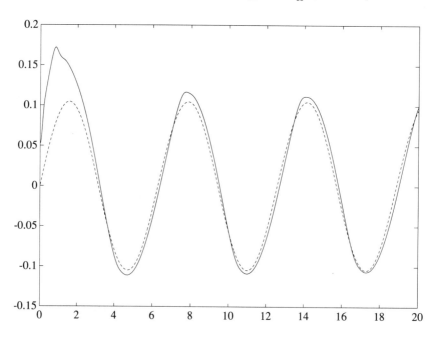

Figure 8.7 The state $x_1(t)$ (solid line) and its desired value $y_m(t) = \frac{\pi}{30}sin(t)$ (dashed line) for the initial condition $\underline{x}(0) = (\frac{\pi}{60}, 0)^T$ in Example 8.1.

Sec. 8.5   Application to Inverted Pendulum Tracking Control   127

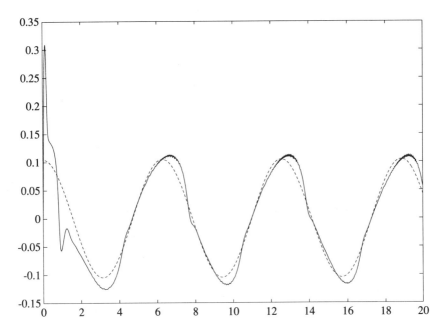

Figure 8.8 The state $x_2(t)$ (solid line) and its desired value $\dot{y}_m(t) = \frac{\pi}{30}cos(t)$ (dashed line) for the initial condition $\underline{x}(0) = (\frac{\pi}{60}, 0)^T$ in Example 8.1.

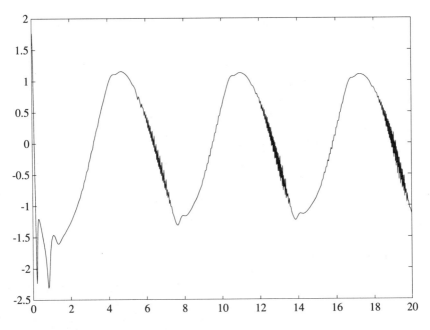

Figure 8.9 The control $u(t)$ for the initial condition $\underline{x}(0) = (\frac{\pi}{60}, 0)^T$ in Example 8.1.

In this case the acceleration of the angle $\theta = x_1$ equals $f(x_1, x_2)$. Based on physical intuition we have the following observation:

$$\text{The bigger the } x_1, \text{ the larger the } f(x_1, x_2) \tag{8.70}$$

Our task now is to transform this fuzzy information into fuzzy rules about $f(x_1, x_2)$. Since $(x_1, x_2) = (0, 0)$ is an (unstable) equilibrium point of the system, we have the first rule:

$$R_f^{(1)} : IF \ x_1 \ is \ F_1^3 \ and \ x_2 \ is \ F_2^3, \ THEN \ f(x_1, x_2) \ is \ near \ zero \tag{8.71}$$

where F_i^j ($i = 1, 2, j = 1, 2, \ldots, 5$) are the fuzzy sets defined in Example 8.1, and "near zero" is a fuzzy set with center zero (since only the centers of the THEN part fuzzy systems, we do not need to specify the detailed membership functions of these THEN part fuzzy sets; that is, knowing their centers is sufficient). From Figure 8.3 we see that the acceleration of x_1 is proportional to the gravity $mg\sin(x_1)$; that is, we have approximately that $f(x_1, x_2) = \alpha \sin(x_1)$, where α is a constant. Clearly, $f(x_1, x_2)$ achieves its maximum at $x_1 = \pi/2$; thus, based on (8.65) we approximately have $a = 16$. Therefore, we have the following fuzzy rules for $f(x_1, x_2)$:

$$R_f^{(2)} : IF \ x_1 \ is \ F_1^1 \ and \ x_2 \ is \ F_2^3, \ THEN \ f(x_1, x_2) \ is \ near \ -8 \tag{8.72}$$

$$R_f^{(3)} : IF \ x_1 \ is \ F_1^2 \ and \ x_2 \ is \ F_2^3, \ THEN \ f(x_1, x_2) \ is \ near \ -4 \tag{8.73}$$

$$R_f^{(4)} : IF \ x_1 \ is \ F_1^4 \ and \ x_2 \ is \ F_2^3, \ THEN \ f(x_1, x_2) \ is \ near \ 4 \tag{8.74}$$

$$R_f^{(5)} : IF \ x_1 \ is \ F_1^5 \ and \ x_2 \ is \ F_2^3, \ THEN \ f(x_1, x_2) \ is \ near \ 8 \tag{8.75}$$

where F_i^j ($i = 1, 2, j = 1, 2, \ldots, 5$) are the fuzzy sets in Example 8.1, and the values in *near* are determined according to $16\sin(\pi/6) = 8$ and $8\sin(\pi/12) \doteq 4$. Also based on physical intuition we have that the $f(x_1, x_2)$ is more sensitive to x_1 than to x_2, we therefore extend the rules (8.71)–(8.75) to the rules where x_2 is any F_2^j for $j = 1, 2, \ldots, 5$. In summary, the final rules characterizing $f(x_1, x_2)$ are shown in Figure 8.10, which comprises 25 rules.

Next, we determine fuzzy rules for $g(x_1, x_2)$ based on physical intuition. Since $g(x_1, x_2)$ determines the strength of the control u on the system and clearly this strength is maximized at $x_1 = 0$, we have the following observation:

$$\text{The smaller the } x_1, \text{ the larger the } g(x_1, x_2) \tag{8.76}$$

Similar to the way of obtaining the rules for $f(x_1, x_2)$ and based on the bounds (8.66) and (8.67), we quantify the observation (8.76) into 25 fuzzy rules for $g(x_1, x_2)$, which are shown in Figure 8.11.

Figures 8.12–8.14 and 8.15–8.17 show the simulation results of the first-type indirect adaptive fuzzy controller for initial conditions $\underline{x}(0) = (-\frac{\pi}{60}, 0)^T$ and $\underline{x}(0) = (\frac{\pi}{60}, 0)^T$, respectively, after the fuzzy rules in Figures 8.10 and 8.11 are incorporated. Comparing these results with those in Example 8.1, we see that the initial parts of control are apparently improved after incorporating these fuzzy rules.

Sec. 8.5 Application to Inverted Pendulum Tracking Control   **129**

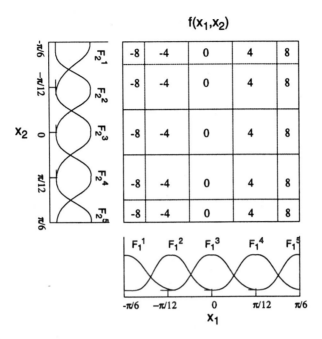

Figure 8.10 Linguistic fuzzy rules for $f(x_1, x_2)$.

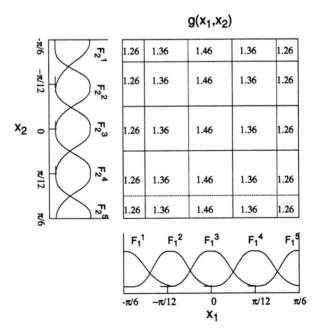

Figure 8.11 Linguistic fuzzy rules for $g(x_1, x_2)$.

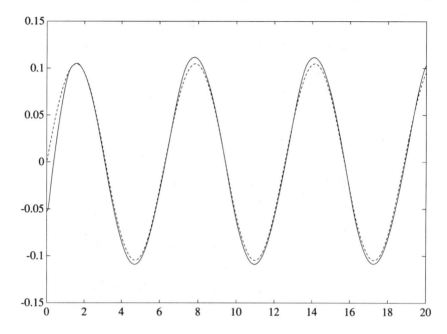

Figure 8.12 The state $x_1(t)$ (solid line) and its desired value $y_m(t) = \frac{\pi}{30}sin(t)$ (dashed line) for the initial condition $\underline{x}(0) = (-\frac{\pi}{60}, 0)^T$ in Example 8.2.

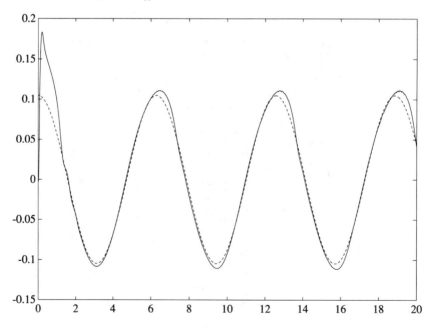

Figure 8.13 The state $x_2(t)$ (solid line) and its desired value $\dot{y}_m(t) = \frac{\pi}{30}cos(t)$ (dashed line) for the initial condition $\underline{x}(0) = (-\frac{\pi}{60}, 0)^T$ in Example 8.2.

Sec. 8.5 Application to Inverted Pendulum Tracking Control

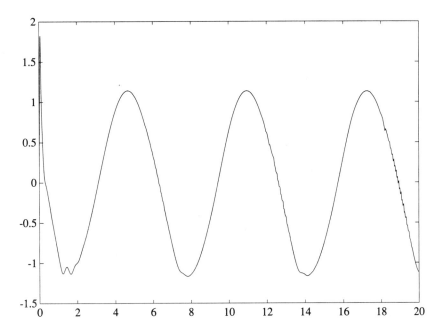

Figure 8.14 The control $u(t)$ for the initial condition $\underline{x}(0) = (-\frac{\pi}{60}, 0)^T$ in Example 8.2.

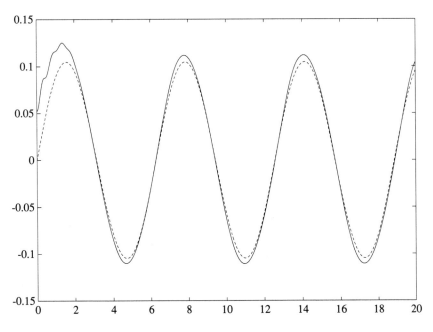

Figure 8.15 The state $x_1(t)$ (solid line) and its desired value $y_m(t) = \frac{\pi}{30} sin(t)$ (dashed line) for the initial condition $\underline{x}(0) = (\frac{\pi}{60}, 0)^T$ in Example 8.2.

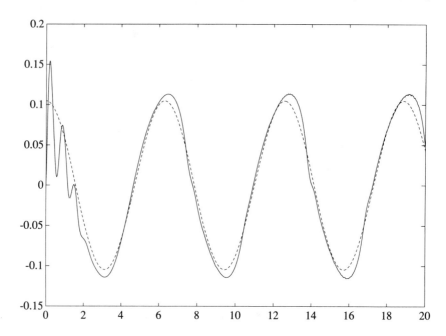

Figure 8.16 The state $x_2(t)$ (solid line) and its desired value $\dot{y}_m(t) = \frac{\pi}{30}cos(t)$ (dashed line) for the initial condition $\underline{x}(0) = (\frac{\pi}{60}, 0)^T$ in Example 8.2.

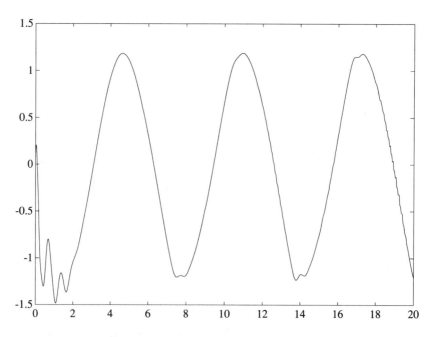

Figure 8.17 The control $u(t)$ for the initial condition $\underline{x}(0) = (\frac{\pi}{60}, 0)^T$ in Example 8.2.

Sec. 8.5 Application to Inverted Pendulum Tracking Control

Example 8.3. In this example we simulated the second-type indirect adaptive fuzzy controller without using any linguistic rules. We chose $M = 15$, the initial $\bar{x}_{fi}^l(0)$ and $\bar{x}_{gi}^l(0)$ randomly in the interval $[-\frac{\pi}{6}, \frac{\pi}{6}]$, the initial $\bar{y}_f^l(0)$ randomly in $[-3, 3]$, the initial $\bar{y}_g^l(0)$ randomly in $[1, 1.3]$, and all the initial σ_{fi}^l and σ_{gi}^l equal to $[\frac{\pi}{6} - (-\frac{\pi}{6})]/15 = \pi/45$. Figures 8.18–8.20 and 8.21–8.23 show the simulation results for initial conditions $\underline{x}(0) = (-\frac{\pi}{60}, 0)^T$ and $\underline{x}(0) = (\frac{\pi}{60}, 0)^T$, respectively, where Figures 8.18 and 8.21 show the state $x_1(t)$ (solid line) and its desired value $y_m(t) = \frac{\pi}{30}sin(t)$ (dashed line), Figures 8.19 and 8.22 show the state $x_2(t)$ (solid line) and its desired value $\dot{y}_m(t) = \frac{\pi}{30}cos(t)$ (dashed line), and Figures 8.20 and 8.23 show the control $u(t)$.

Example 8.4. Here we considered the same situation as in Example 8.3 except that we used the linguistic rules in Figures 8.10 and 8.11. Since Figures 8.10 and 8.11 each has 25 rules while the fuzzy systems in the controller are constructed each by 15 rules, we only used the two sets of 15 rules corresponding to the middle three columns of Figures 8.10 and 8.11; that is, we used these $15 + 15$ rules to construct the initial controller. Figures 8.24–8.26 and 8.27–8.29 show the simulation results in this case for initial conditions $\underline{x}(0) = (-\frac{\pi}{60}, 0)^T$ and $\underline{x}(0) = (\frac{\pi}{60}, 0)^T$, respectively. Comparing these results with those in Example 8.3, we see that the control performance is apparently improved after incorporating these fuzzy rules.

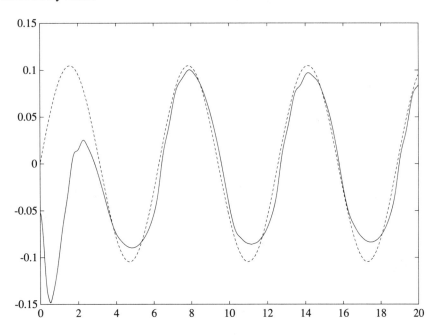

Figure 8.18 The state $x_1(t)$ (solid line) and its desired value $y_m(t) = \frac{\pi}{30}sin(t)$ (dashed line) for the initial condition $\underline{x}(0) = (-\frac{\pi}{60}, 0)^T$ in Example 8.3.

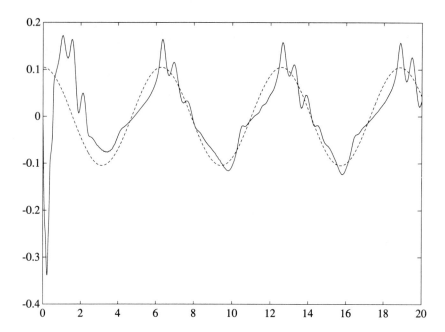

Figure 8.19 The state $x_2(t)$ (solid line) and its desired value $\dot{y}_m(t) = \frac{\pi}{30}cos(t)$ (dashed line) for the initial condition $\underline{x}(0) = (-\frac{\pi}{60}, 0)^T$ in Example 8.3.

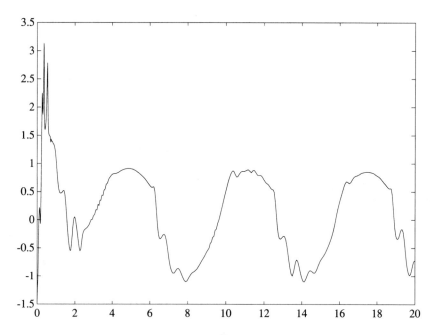

Figure 8.20 The control $u(t)$ for the initial condition $\underline{x}(0) = (-\frac{\pi}{60}, 0)^T$ in Example 8.3.

Sec. 8.5 Application to Inverted Pendulum Tracking Control

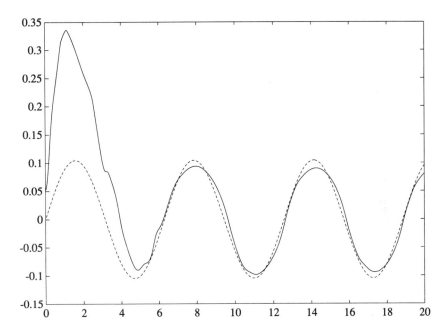

Figure 8.21 The state $x_1(t)$ (solid line) and its desired value $y_m(t) = \frac{\pi}{30}sin(t)$ (dashed line) for the initial condition $\underline{x}(0) = (\frac{\pi}{60}, 0)^T$ in Example 8.3.

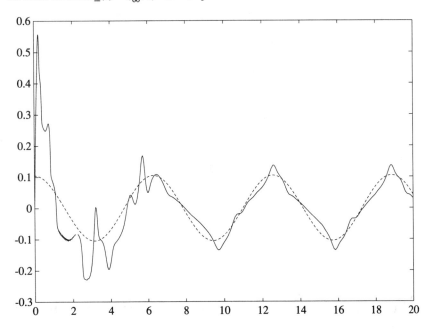

Figure 8.22 The state $x_2(t)$ (solid line) and its desired value $\dot{y}_m(t) = \frac{\pi}{30}cos(t)$ (dashed line) for the initial condition $\underline{x}(0) = (\frac{\pi}{60}, 0)^T$ in Example 8.3.

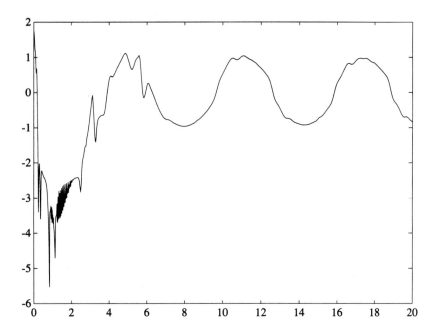

Figure 8.23 The control $u(t)$ for the initial condition $\underline{x}(0) = (\frac{\pi}{60}, 0)^T$ in Example 8.3.

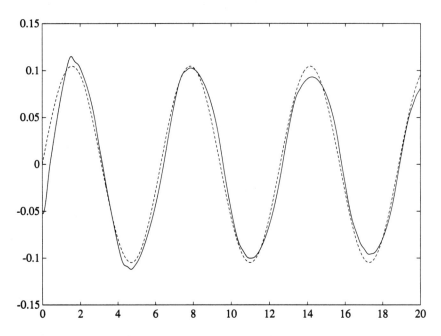

Figure 8.24 The state $x_1(t)$ (solid line) and its desired value $y_m(t) = \frac{\pi}{30}sin(t)$ (dashed line) for the initial condition $\underline{x}(0) = (-\frac{\pi}{60}, 0)^T$ in Example 8.4.

Sec. 8.5 Application to Inverted Pendulum Tracking Control

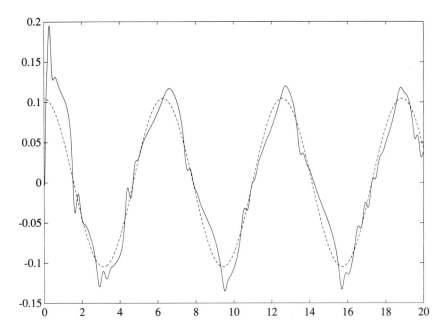

Figure 8.25 The state $x_2(t)$ (solid line) and its desired value $\dot{y}_m(t) = \frac{\pi}{30} cos(t)$ (dashed line) for the initial condition $\underline{x}(0) = (-\frac{\pi}{60}, 0)^T$ in Example 8.4.

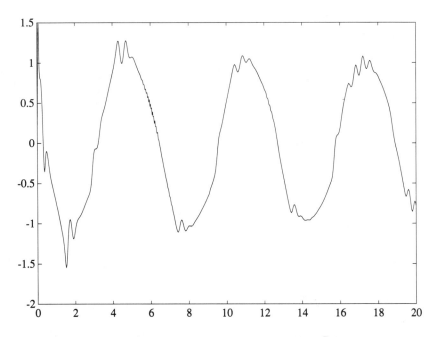

Figure 8.26 The control $u(t)$ for the initial condition $\underline{x}(0) = (-\frac{\pi}{60}, 0)^T$ in Example 8.4.

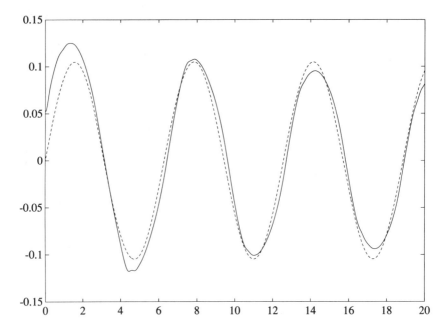

Figure 8.27 The state $x_1(t)$ (solid line) and its desired value $y_m(t) = \frac{\pi}{30}sin(t)$ (dashed line) for the initial condition $\underline{x}(0) = (\frac{\pi}{60}, 0)^T$ in Example 8.4.

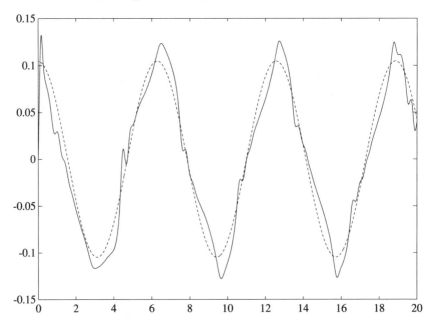

Figure 8.28 The state $x_2(t)$ (solid line) and its desired value $\dot{y}_m(t) = \frac{\pi}{30}cos(t)$ (dashed line) for the initial condition $\underline{x}(0) = (\frac{\pi}{60}, 0)^T$ in Example 8.4.

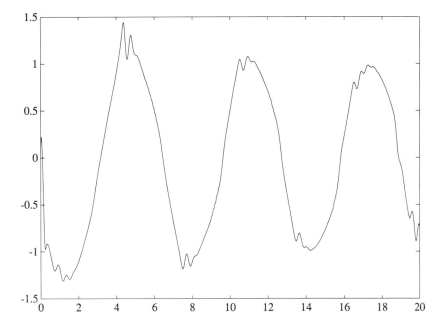

Figure 8.29 The control $u(t)$ for the initial condition $\underline{x}(0) = (\frac{\pi}{60}, 0)^T$ in Example 8.4.

8.6 CONCLUDING REMARKS

In this chapter, we developed two indirect adaptive fuzzy controllers which (1) do not require an accurate mathematical model of the system under control, (2) are capable of incorporating fuzzy IF-THEN rules describing the system directly into the controllers, and (3) guarantee the global stability of the resulting closed-loop systems in the sense that all signals involved are uniformly bounded. We provide the specific formula of the bounds so that controller designers can determine the bounds based on their requirements. We used the adaptive fuzzy controllers to control the inverted pendulum to track a sine-wave trajectory, and the simulation results show that (1) the adaptive fuzzy controllers could perform successful tracking without using any linguistic information, and (2) after incorporating some linguistic fuzzy rules into the controllers, the adaptation speed became faster and the tracking error became smaller.

9

STABLE DIRECT ADAPTIVE FUZZY CONTROL OF NONLINEAR SYSTEMS

9.1 INTRODUCTION

In this chapter we construct direct adaptive fuzzy controllers. As described in Section 8.1.3, direct adaptive fuzzy controllers use adaptive fuzzy systems directly as controllers and should be able to incorporate fuzzy control rules.

In Section 9.2, we show the basic ideas of constructing direct adaptive fuzzy controllers. In Sections 9.3 and 9.4, we present the detailed steps of designing the first and second types of direct adaptive fuzzy controllers, respectively, and study their properties. In Section 9.5, we show simulation results.

9.2 BASIC IDEAS OF CONSTRUCTING DIRECT ADAPTIVE FUZZY CONTROLLERS

The approach of this section is very similar to that of Section 8.2. Consider the system

$$x^{(n)} = f(x, \dot{x}, \ldots, x^{(n-1)}) + bu, \qquad y = x \qquad (9.1)$$

where f is an unknown function, b is a positive unknown constant, and $u \in R$ and $y \in R$ are the input and output of the system, respectively. We assume that the state vector $\underline{x} = (x, \dot{x}, \ldots, x^{(n-1)})^T$ is available for measurement. The control objectives are the same as those in Section 8.2; the only difference is that the basic control $u_c(\underline{x}|\underline{\theta})$ must be a fuzzy logic system.

Sec. 9.2 Basic Ideas of Constructing Direct Adaptive Fuzzy Controllers

Suppose that the control u is the summation of a basic control $u_c(\underline{x}|\underline{\theta})$ and a supervisory control $u_s(\underline{x})$:

$$u = u_c(\underline{x}|\underline{\theta}) + u_s(\underline{x}) \tag{9.2}$$

where $u_c(\underline{x}|\underline{\theta})$ is a fuzzy logic system which is in the form of either (8.1) or (8.3), and $u_s(\underline{x})$ will be determined later in this section. Substituting (9.2) into (9.1), we have

$$x^{(n)} = f(\underline{x}) + b[u_c(\underline{x}|\underline{\theta}) + u_s(\underline{x})] \tag{9.3}$$

If $f(\underline{x})$ and b are known, then from Section 8.2 we know that the control

$$u^* = \frac{1}{b}[-f + y_m^{(n)} + \underline{k}^T \underline{e}] \tag{9.4}$$

will force the $\underline{e}$ to converge to zero, where $\underline{e} = (e, \dot{e}, \ldots, e^{(n-1)})^T$ and $\underline{k} = (k_n, \ldots, k_1)^T$ such that all roots of $s^n + k_1 s^{n-1} + \cdots + k_n = 0$ are in the open left half-plane. Now adding and substracting bu^* to (9.3) and after some straightforward manipulation, we obtain the error equation governing the closed-loop system:

$$e^{(n)} = -\underline{k}^T \underline{e} + b[u^* - u_c(\underline{x}|\underline{\theta}) - u_s(\underline{x})] \tag{9.5}$$

or equivalently,

$$\dot{\underline{e}} = \Lambda_c \underline{e} + \underline{b}_c[u^* - u_c(\underline{x}|\underline{\theta}) - u_s(\underline{x})] \tag{9.6}$$

where

$$\Lambda_c = \begin{bmatrix} 0 & 1 & 0 & 0 & \cdots & 0 & 0 \\ 0 & 0 & 1 & 0 & \cdots & 0 & 0 \\ \cdots & \cdots & \cdots & \cdots & \cdots & & \\ 0 & 0 & 0 & 0 & \cdots & 0 & 1 \\ -k_n & -k_{n-1} & \cdots & \cdots & \cdots & & -k_1 \end{bmatrix}, \underline{b}_c = \begin{bmatrix} 0 \\ \cdots \\ 0 \\ b \end{bmatrix} \tag{9.7}$$

Similar to Section 8.2, define $V_e = \frac{1}{2}\underline{e}^T P \underline{e}$, where P is a symmetric positive definite matrix satisfying the Lyapunov equation

$$\Lambda_c^T P + P \Lambda_c = -Q \tag{9.8}$$

where $Q > 0$. Using (9.8) and the error equation (9.6), we have

$$\dot{V}_e = -\frac{1}{2}\underline{e}^T Q \underline{e} + \underline{e}^T P \underline{b}_c[u^* - u_c(\underline{x}|\underline{\theta}) - u_s(\underline{x})]$$

$$\leq -\frac{1}{2}\underline{e}^T Q \underline{e} + |\underline{e}^T P \underline{b}_c|(|u^*| + |u_c|) - \underline{e}^T P \underline{b}_c u_s \tag{9.9}$$

In order to design the u_s such that $\dot{V}_e \leq 0$, we need the following assumption:

Assumption 9.1. We can determine a function $f^U(\underline{x})$ and a constant b_L such that $|f(\underline{x})| \leq f^U(\underline{x})$ and $0 \leq b_L \leq b$.

We construct the supervisory control $u_s(\underline{x})$ as follows:

$$u_s(\underline{x}) = I_1^* sgn(\underline{e}^T P \underline{b}_c)[|u_c| + \frac{1}{b_L}(f^U + |y_m^{(n)}| + |\underline{k}^T \underline{e}|)] \tag{9.10}$$

where $I_1^* = 1$ if $V_e > \bar{V}$ (which is a constant specified by the designer), and $I_1^* = 0$ if $V_e \leq \bar{V}$. Because $b > 0$, $sgn(\underline{e}^T P \underline{b}_c)$ can be determined; also, all other terms in (9.10) can be determined, thus the supervisory control u_s of (9.10) can be implemented. Because of I_1^*, the u_s is a supervisory kind of controller. Substituting (9.10) and (9.4) into (9.9) and considering the $I_1^* = 1$ case, we have

$$\dot{V}_e \leq -\frac{1}{2}\underline{e}^T Q\underline{e} + |\underline{e}^T P\underline{b}_c| \left[\frac{1}{b}\left(|f| + |y_m^{(n)}| + |\underline{k}^T \underline{e}|\right) \right.$$
$$\left. + |u_c| - |u_c| - \frac{1}{b_L}\left(f^U + |y_m^{(n)}| + |\underline{k}^T \underline{e}|\right) \right]$$
$$\leq -\frac{1}{2}\underline{e}^T Q\underline{e} \leq 0 \qquad (9.11)$$

Therefore, using the supervisory control u_s of (9.10), we always have $V_e \leq \bar{V}$. Because $P > 0$, the boundedness of V_e implies the boundedness of $\underline{e}$, which in turn implies the boundedness of $\underline{x}$.

Next we replace the $u_c(\underline{x}|\underline{\theta})$ by the fuzzy logic system (8.1) or (8.3) and develop an adaptive law to adjust the parameter vector $\underline{\theta}$. Define the optimal parameter vector:

$$\underline{\theta}^* \equiv argmin_{|\underline{\theta}| \leq M_\theta} [sup_{|\underline{x}| \leq M_x} |u_c(\underline{x}|\underline{\theta}) - u^*|] \qquad (9.12)$$

and the minimum approximation error:

$$w \equiv u_c(\underline{x}|\underline{\theta}^*) - u^* \qquad (9.13)$$

The error equation (9.6) can be rewritten as

$$\dot{\underline{e}} = \Lambda_c \underline{e} + \underline{b}_c [u_c(\underline{x}|\underline{\theta}^*) - u_c(\underline{x}|\underline{\theta})] - \underline{b}_c u_s(\underline{x}) - \underline{b}_c w \qquad (9.14)$$

If we choose the $u_c(\underline{x}|\underline{\theta})$ to be in the form of (8.1), then (9.14) becomes

$$\dot{\underline{e}} = \Lambda_c \underline{e} + \underline{b}_c \underline{\phi}^T \underline{\xi}(\underline{x}) - \underline{b}_c u_s - \underline{b}_c w \qquad (9.15)$$

where $\underline{\phi} \equiv \underline{\theta}^* - \underline{\theta}$ and $\underline{\xi}(\underline{x})$ is the fuzzy basis function (8.2). Define the Lyapunov function candidate

$$V = \frac{1}{2}\underline{e}^T P\underline{e} + \frac{b}{2\gamma}\underline{\phi}^T \underline{\phi} \qquad (9.16)$$

Using (9.15) and (9.8), we have

$$\dot{V} = -\frac{1}{2}\underline{e}^T Q\underline{e} + \underline{e}^T P\underline{b}_c(\underline{\phi}^T \underline{\xi}(\underline{x}) - u_s - w) + \frac{b}{\gamma}\underline{\phi}^T \dot{\underline{\phi}} \qquad (9.17)$$

Let $\underline{p}_n$ be the last column of P, then from (9.7) we have

$$\underline{e}^T P\underline{b}_c = \underline{e}^T \underline{p}_n b \qquad (9.18)$$

Substituting (9.18) into (9.17), we have

$$\dot{V} = -\frac{1}{2}\underline{e}^T Q\underline{e} + \frac{b}{\gamma}\underline{\phi}^T (\gamma \underline{e}^T \underline{p}_n \underline{\xi}(\underline{x}) + \dot{\underline{\phi}}) - \underline{e}^T P\underline{b}_c u_s - \underline{e}^T P\underline{b}_c w \qquad (9.19)$$

Sec. 9.2 Basic Ideas of Constructing Direct Adaptive Fuzzy Controllers

If we choose the adaptive law:

$$\dot{\underline{\theta}} = \gamma \underline{e}^T \underline{p}_n \underline{\xi}(\underline{x}) \quad (9.20)$$

then (9.19) becomes

$$\dot{V} \leq -\frac{1}{2}\underline{e}^T Q \underline{e} - \underline{e}^T P \underline{b}_c w \quad (9.21)$$

where we use the facts $\underline{e}^T P \underline{b}_c u_s \geq 0$ (observe (9.10)) and $\dot{\underline{\phi}} = -\dot{\underline{\theta}}$. As in Section 8.2, this is the best we can achieve. In order to guarantee $|\underline{\theta}| \leq M_\theta$, we use the projection algorithm to modify the basic adaptive law (9.20).

If we choose $u_c(\underline{x}|\underline{\theta})$ to be the fuzzy logic system (8.3), then we take the Taylor series expansion of $u_c(\underline{x}|\underline{\theta}^*)$ around $\underline{\theta}$:

$$u_c(\underline{x}|\underline{\theta}^*) - u_c(\underline{x}|\underline{\theta}) = \underline{\phi}^T \frac{\partial u_c}{\partial \underline{\theta}} + O(|\underline{\phi}|^2) \quad (9.22)$$

Substituting (9.22) into (9.14), we obtain the error equation for this case:

$$\dot{\underline{e}} = \Lambda_c \underline{e} + \underline{b}_c \underline{\phi}^T \frac{\partial u_c}{\partial \underline{\theta}} - \underline{b}_c u_s - \underline{b}_c v \quad (9.23)$$

where $v \equiv w + O(|\underline{\phi}|^2)$. The rest of the development is the same as (9.16)–(9.21); we omit the details. The overall control scheme is shown in Figure 9.1.

Finally, we specify the linguistic information which can be directly incorporated into the direct adaptive fuzzy controllers developed in this section.

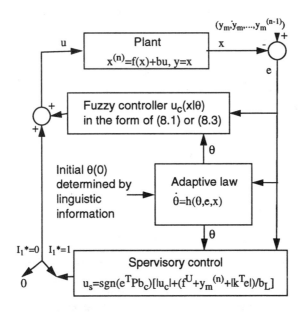

Figure 9.1 The overall scheme of direct adaptive fuzzy control.

Assumption 9.2. There are L fuzzy control rules from human experts in the following form:

$$R_c^{(r)} : IF \ x_1 \ is \ A_1^r \ and \ \cdots \ and \ x_n \ is \ A_n^r, \ THEN \ u \ is \ C^r \quad (9.24)$$

where A_i^r and C^r are labels of fuzzy sets in R, and $r = 1, 2, \ldots, L$. Again, we allow $L = 0$ which means that there is no linguistic information in the form of (9.24).

9.3 DESIGN AND STABILITY ANALYSIS OF FIRST-TYPE DIRECT ADAPTIVE FUZZY CONTROLLERS

In this section, we choose $u_c(\underline{x}|\underline{\theta})$ to be the fuzzy logic system in the form of (8.1). We first present the detailed design steps of the adaptive fuzzy controller and then study its properties.

Design of First-Type Direct Adaptive Fuzzy Controllers

Step 1: Off-line Preprocessing

- Specify the $k_1, \ldots, k_n$ such that all roots of $s^n + k_1 s^{n-1} + \cdots + k_n = 0$ are in the open left half-plane. Specify a positive definite $n \times n$ matrix Q.
- Solve the Lyapunov equation (9.8), for example, by using the method in Wang and Mendel [90], to obtain a symmetric $P > 0$.
- Specify the design parameters M_θ and M_x based on practical constraints.

Step 2: Initial Controller Construction

- Define m_i fuzzy sets $F_i^{l_i}$ whose membership functions $\mu_{F_i^{l_i}}$ uniformly cover U_{c_i} which is the projection of U_c onto the ith coordinate, where $l_i = 1, 2, \ldots, m_i$ and $i = 1, 2, \ldots, n$. The U_c is usually chosen to be $\{\underline{x} \in R^n : |\underline{x}| \leq M_x\}$. We require that the $F_i^{l_i}$'s include the A_i^r's in (9.24).
- Construct the fuzzy rule base for the fuzzy logic system $u_c(\underline{x}|\underline{\theta})$ which consists of $m_1 \times m_2 \times \cdots \times m_n$ rules whose IF parts comprise all the possible combinations of the $F_i^{l_i}$'s for $i = 1, 2, \ldots, n$. Specifically, the fuzzy rule base of $u_c(\underline{x}|\underline{\theta})$ consists of rules:

$$R_c^{(l_1,\ldots,l_n)} : \quad IF \ x_1 \ is \ F_1^{l_1} \ and \ \cdots \ and \ x_n \ is \ F_n^{l_n},$$
$$THEN \ u_c \ is \ G^{(l_1,\ldots,l_n)} \quad (9.25)$$

where $l_i = 1, 2, \ldots, m_i$, $i = 1, 2, \ldots, n$, and $G^{(l_1,\ldots,l_n)}$ are fuzzy sets in R which are specified as follows: if the IF part of (9.25) agrees with the IF part of (9.24), set $G^{(l_1,\ldots,l_n)}$ equal to the corresponding C^r; otherwise, set $G^{(l_1,\ldots,l_n)}$ arbitrarily with the constraint that the centers of $G^{(l_1,\ldots,l_n)}$ (which correspond to the $\bar{y}^l$ parameters) are inside the constraint sets $\{\underline{\theta} : |\underline{\theta}| \leq M_\theta\}$. Therefore, *the initial adaptive fuzzy controller is constructed from the fuzzy control rules* (9.24).

Sec. 9.3 Design and Stability Analysis of First-Type Direct Adaptive Fuzzy Controllers

- Construct the fuzzy basis functions

$$\xi^{(l_1,\ldots,l_n)}(\underline{x}) = \frac{\prod_{i=1}^n \mu_{F_i^{l_i}}(x_i)}{\sum_{l_1=1}^{m_1} \cdots \sum_{l_n=1}^{m_n}(\prod_{i=1}^n \mu_{F_i^{l_i}}(x_i))} \quad (9.26)$$

and collect them into a $\prod_{i=1}^n m_i$-dimensional vector $\underline{\xi}(\underline{x})$ in a natural ordering for $l_1 = 1, 2, \ldots, m_1, \ldots, l_n = 1, 2, \ldots, m_n$. Collect the points at which $\mu_{G^{(l_1,\ldots,l_n)}}$'s achieve their maximum values, in the same ordering as $\underline{\xi}(\underline{x})$, into the vector $\underline{\theta}$. The $u_c(\underline{x}|\underline{\theta})$ is constructed as

$$u_c(\underline{x}|\underline{\theta}) = \underline{\theta}^T \underline{\xi}(\underline{x}) \quad (9.27)$$

Step 3: On-line Adaptation

- Apply the feedback control (9.2) to the plant (9.1), where u_c is given by (9.27), and u_s is given by (9.10).
- Use the following adaptive law to adjust the parameter vector $\underline{\theta}$:

$$\underline{\dot{\theta}} = \begin{cases} \gamma \underline{e}^T \underline{p}_n \underline{\xi}(\underline{x}) & \text{if } (|\underline{\theta}| < M_\theta) \text{ or } (|\underline{\theta}| = M_\theta \text{ and } \underline{e}^T \underline{p}_n \underline{\theta}^T \underline{\xi}(\underline{x}) \geq 0) \\ P\{\gamma \underline{e}^T \underline{p}_n \underline{\xi}(\underline{x})\} & \text{if } (|\underline{\theta}| = M_\theta \text{ and } \underline{e}^T \underline{p}_n \underline{\theta}^T \underline{\xi}(\underline{x}) < 0) \end{cases} \quad (9.28)$$

where the projection operator $P\{*\}$ is defined as:

$$P\{\gamma \underline{e}^T \underline{p}_n \underline{\xi}(\underline{x})\} = \gamma \underline{e}^T \underline{p}_n \underline{\xi}(\underline{x}) - \gamma \underline{e}^T \underline{p}_n \frac{\underline{\theta} \underline{\theta}^T \underline{\xi}(\underline{x})}{|\underline{\theta}|^2} \quad (9.29)$$

The following theorem shows the properties of this adaptive fuzzy controller.

Theorem 9.1. Consider the plant (9.1) with the control (9.2), where u_c is given by (9.27), and u_s is given by (9.10). Let the parameter vector $\underline{\theta}$ be adjusted by the adaptive law (9.28), and let Assumptions 9.1 and 9.2 be true. Then, the overall control scheme (shown in Figure 9.1) guarantees the following properties:

1. $|\underline{\theta}(t)| \leq M_\theta$,

$$|\underline{x}(t)| \leq |\underline{y}_m| + (\frac{2\bar{V}}{\lambda_{min}})^{1/2} \quad (9.30)$$

and

$$|u(t)| \leq 2M_\theta + \frac{1}{b_L}[f^U + |y_m^{(n)}| + |\underline{k}|(\frac{2\bar{V}}{\lambda_{min}})^{1/2}] \quad (9.31)$$

for all $t \geq 0$, where λ_{min} is the minimum eigenvalue of P, and $\underline{y}_m = (y_m, \dot{y}_m, \ldots, y_m^{(n-1)})^T$.

2.

$$\int_0^t |\underline{e}(\tau)|^2 d\tau \leq a + b \int_0^t |w(\tau)|^2 d\tau \qquad (9.32)$$

for all $t \geq 0$, where a and b are constants, and w is the minimum approximation error defined by (9.13).

3. If w is squared integrable, that is, $\int_0^\infty |w(t)|^2 dt < \infty$, then $lim_{t\to\infty} |\underline{e}(t)| = 0$.

Proof of this theorem is given in the Appendix.

Remark 9.1. The points made in Remarks 8.1–8.5 apply to this adaptive fuzzy controller. Indeed, comparing Sections 8.3 with 9.3, we see that they are very similar.

9.4 DESIGN AND STABILITY ANALYSIS OF SECOND-TYPE DIRECT ADAPTIVE FUZZY CONTROLLERS

Design of Second-Type Direct Adaptive Fuzzy Controllers

Step 1: Off-line Preprocessing

This is the same as the first-type direct adaptive fuzzy controller, except that we need to specify one more design parameter σ.

Step 2: Initial Controller Construction

- Choose $u_c(\underline{x}|\underline{\theta})$ to be the fuzzy logic system in the form of (8.3); that is,

$$u_c(\underline{x}|\underline{\theta}) = \frac{\sum_{l=1}^M \bar{y}^l \left[\prod_{i=1}^n exp\left(-\left(\frac{x_i - \bar{x}_i^l}{\sigma_i^l}\right)^2\right)\right]}{\sum_{l=1}^M \left[\prod_{i=1}^n exp\left(-\left(\frac{x_i - \bar{x}_i^l}{\sigma_i^l}\right)^2\right)\right]} \qquad (9.33)$$

where $\underline{\theta}$ is the collection of the adjustable parameters $\bar{y}^l, \bar{x}_i^l$, and σ_i^l. Clearly, u_c in (9.33) is constructed based on M rules whose IF parts are characterized by the Gaussian membership functions

$$\mu_{F_i^l}(x_i) = exp\left(-\left(\frac{x_i - \bar{x}_i^l}{\sigma_i^l}\right)^2\right) \qquad (9.34)$$

where $l = 1, 2, \ldots, M$ and $i = 1, 2, \ldots, n$. For this adaptive fuzzy controller, we assume that the A_i^r in (9.24) are also characterized by the Gaussian membership functions in the form of (9.34), and that $L \leq M$.

- Determine the initial $\underline{\theta}(0)$ as follows: for the L $\mu_{F_i^l}$'s that are the same as the $\mu_{A_i^r}$'s in (9.24), choose $\bar{x}_i^l(0)$ and $\sigma_i^l(0)$ to be the same as in the $\mu_{A_i^r}$'s, and choose the $\bar{y}^l(0)$'s to be the centers of the corresponding μ_{C^r}'s; the remaining parameters are chosen arbitrarily in the constraint sets $\{\underline{\theta} : |\underline{\theta}| \leq M_\theta, \sigma_i^l \geq \sigma\}$.

Design and Stability Analysis of Second-Type Direct Adaptive Fuzzy Controllers

Step 3: On-line Adaptation

- Compute $\frac{\partial u_c(\underline{x}|\underline{\theta})}{\partial \underline{\theta}}$ using the following algorithm:

$$\frac{\partial u_c}{\partial \bar{y}^l} = \frac{b^l}{\sum_{l=1}^{M} b^l} \tag{9.35}$$

$$\frac{\partial u_c}{\partial \bar{x}_i^l} = \frac{\bar{y}^l - u_c}{\sum_{l=1}^{M} b^l} b^l \frac{-2(x_i - \bar{x}_i^l)}{(\sigma_i^l)^2} \tag{9.36}$$

$$\frac{\partial u_c}{\partial \sigma_i^l} = \frac{\bar{y}^l - u_c}{\sum_{l=1}^{M} b^l} b^l \frac{-2(x_i - \bar{x}_i^l)^2}{(\sigma_i^l)^3} \tag{9.37}$$

where

$$b^l = \prod_{i=1}^{n} exp\left(-\left(\frac{x_i - \bar{x}_i^l}{\sigma_i^l}\right)^2\right) \tag{9.38}$$

The preceding equations are obtained by taking differentials of u_c in the form of (9.33) with respect to the corresponding parameters.

- Apply the feedback control (9.2) to the plant (9.1), where u_c is given by (9.33), and u_s is given by (9.10).
- Use the following adaptive law to adjust the parameter vector $\underline{\theta}$:
  □ Whenever $\sigma_i^l = \sigma$, use

$$\dot{\sigma}_i^l = \begin{cases} \gamma \underline{e}^T \underline{p}_n \frac{\partial u_c}{\partial \sigma_i^l} & if \quad \underline{e}^T \underline{p}_n \frac{\partial u_c}{\partial \sigma_i^l} < 0 \\ \\ 0 & if \quad \underline{e}^T \underline{p}_n \frac{\partial u_c}{\partial \sigma_i^l} \geq 0 \end{cases} \tag{9.39}$$

  □ Otherwise, use

$$\dot{\underline{\theta}} = \begin{cases} \gamma \underline{e}^T \underline{p}_n \frac{\partial u_c}{\partial \underline{\theta}} & if \quad (|\underline{\theta}| < M_\theta) \quad or \quad (|\underline{\theta}| = M_\theta \quad and \quad \underline{e}^T \underline{p}_n \underline{\theta}^T \frac{\partial u_c}{\partial \underline{\theta}} \geq 0) \\ \\ \gamma \underline{e}^T \underline{p}_n \frac{\partial u_c}{\partial \underline{\theta}} - \gamma \underline{e}^T \underline{p}_n \frac{\underline{\theta}\underline{\theta}^T}{|\underline{\theta}|^2} \frac{\partial u_c}{\partial \underline{\theta}} & if \quad (|\underline{\theta}| = M_\theta \quad and \quad \underline{e}^T \underline{p}_n \underline{\theta}^T \frac{\partial u_c}{\partial \underline{\theta}} < 0) \end{cases} \tag{9.40}$$

Properties of this second-type direct adaptive fuzzy controller are summarized in the following theorem.

Theorem 9.2. The second-type direct adaptive fuzzy controller designed through the preceding three steps guarantees the following properties:

1. $|\underline{\theta}| \leq M_\theta$, the σ_i^l's in $\underline{\theta} \geq \sigma$, and $\underline{x}$ and u satisfy (9.30) and (9.31), respectively.

2.

$$\int_0^t |\underline{e}(\tau)|^2 d\tau \le a + b \int_0^t |v(\tau)|^2 d\tau \tag{9.41}$$

for all $t \ge 0$, where a and b are constants, and $v = w + O(|\underline{\theta}|^2)$.

3. If v is squared integrable, then $lim_{t\to\infty}|\underline{e}(t)| = 0$.

Proof of this theorem is given in the Appendix.

Remark 9.2. Remarks 8.1–8.5 and 8.8 apply to this adaptive fuzzy controller.

9.5 SIMULATIONS

Example 9.1. In this example, we use the first-type direct adaptive fuzzy controller to regulate the plant

$$\dot{x}(t) = \frac{1 - e^{-x(t)}}{1 + e^{-x(t)}} + u(t) \tag{9.42}$$

to the origin; that is, $y_m \equiv 0$. It is clear that the plant (9.42) is unstable if without control because if $u(t) \equiv 0$, then $\dot{x} = \frac{1-e^{-x}}{1+e^{-x}} > 0$ for $x > 0$, and $\dot{x} = \frac{1-e^{-x}}{1+e^{-x}} < 0$ for $x < 0$. We choose $\gamma = 1$, $M_x = 3$, $M_\theta = 3$, $b_L = 0.5 < 1 = b$, and $f^U = 1 \ge \frac{1-e^{-x(t)}}{1+e^{-x(t)}}$. In Step 1, we define six fuzzy sets over the interval $[-3, 3]$, with labels $N3, N2, N1, P1, P2$, and $P3$, and membership functions $\mu_{N3}(x) = 1/(1 + exp(5(x + 2)))$, $\mu_{N2}(x) = exp(-(x + 1.5)^2)$, $\mu_{N1}(x) = exp(-(x + 0.5)^2)$, $\mu_{P1}(x) = exp(-(x - 0.5)^2)$, $\mu_{P2}(x) = exp(-(x - 1.5)^2)$, and $\mu_{P3}(x) = 1/(1+exp(-5(x - 2)))$, as shown in Figure 9.2. In Step 2, we consider two cases: (1) there are no fuzzy control rules, and the initial $\theta_i(0)$'s are chosen randomly in the interval $[-2, 2]$; and (2) there are two fuzzy control rules:

$$R_1 : \quad IF \ x \ is \ N2, \ THEN \ u(x) \ is \ PB \tag{9.43}$$

$$R_2 : \quad IF \ x \ is \ P2, \ THEN \ u(x) \ is \ NB \tag{9.44}$$

where $\mu_{PB}(u) = exp(-(u-2)^2)$, and $\mu_{NB}(u) = exp(-(u+2)^2)$. These two rules are obtained by considering the fact that our problem is to control $x(t)$ to zero; therefore, if x is negative, then the control $u(x)$ should be *positive big* (PB) so that it may happen that $\dot{x} > 0$ (see (9.42)). On the other hand, if x is positive, then the control $u(x)$ should be *negative big* (NB) so that it may happen that $\dot{x} < 0$. We used MATLAB command "ode23" to simulate the overall control system. We chose the initial state $x(0) = 1$. Figures 9.3 and 9.4 show the $x(t)$ for the cases without and with the linguistic control rules (9.43) and (9.44), respectively. We see from Figures 9.3 and 9.4 that (1) the first-type direct adaptive fuzzy controller could regulate the plant to the origin without using the fuzzy control rules; and (2) by using the fuzzy control rules, the speed of convergence became much faster. We also simulated for other initial conditions, and the results were very similar. We do not show them in order to clear the figures and make the comparisons easier.

Sec. 9.5 Simulations

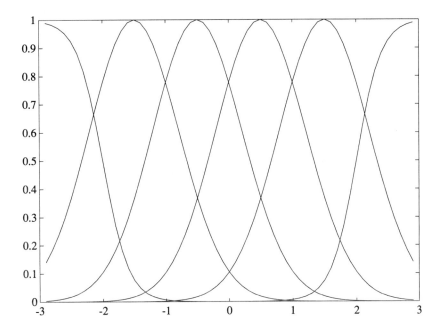

Figure 9.2 Fuzzy membership functions defined over the state space for the example.

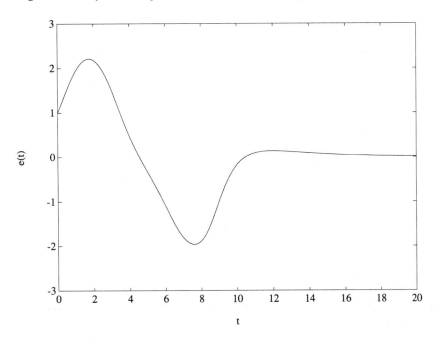

Figure 9.3 Closed-loop system state $x(t) = e(t)$ using the first-type direct adaptive fuzzy controller for the plant (9.42) without incorporating any fuzzy control rules.

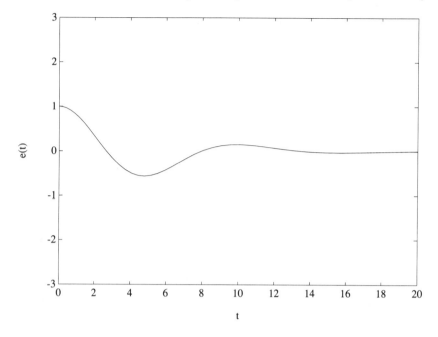

Figure 9.4 Closed-loop system state $x(t) = e(t)$ using the first-type direct adaptive fuzzy controller for the plant (9.42) after incorporating the fuzzy control rules (9.43) and (9.44).

In the simulations shown in Figures 9.3 and 9.4, the state $x(t)$ did not hit the boundary $|x| = 3$; therefore, the supervisory control u_s never fired. Now we change $M_x = 1.5$ and keep all other parameters the same as in the simulation shown in Figure 9.3 (that is, without using the fuzzy control rules). The simulation result for this case is shown in Figure 9.5. We see that the supervisory control u_s did force the state to be inside the constraint set $|x| \leq 1.5$.

It is interesting to observe how this bounded control was achieved. From Figure 9.5 we see that as soon as the state hit the boundary, the supervisory control u_s began operating to force the state back to the constraint set. As soon as the state was inside the constraint set, the supervisory control stopped operating, which caused the state to hit the boundary again. The continuation of this kind of back-and-forth operation resulted in the "holding" of the state around the boundary, as shown in Figure 9.5 in the intervals $t \in [0.5, 2.5]$ and $t \in [5.5, 7]$ (approximately). That is, the supervisory control could prevent the system from being unstable but could not regulate the state to the origin. The encouraging observation is that during these "holding periods," the adaptive law adjusted the parameters of the fuzzy controller u_c, and, finally, the fuzzy control "recovered" and finished the control task—regulating the state to the origin.

We may view the fuzzy controller u_c as a lower-level operator and the supervisory control u_s as a higher-level supervisor. If the operator could perform successful control, the supervisor just observes and does not take any action. If

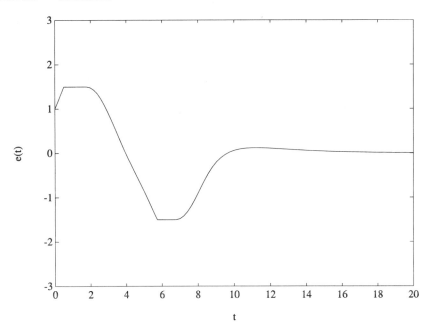

Figure 9.5 Closed-loop system state $x(t) = e(t)$ for the same situation as in Figure 9.3 except $M_x = 1.5$.

somehow the operator could not control the system well, the supervisor takes actions to prevent catastrophic consequences. During this period, the operator learns to correct mistakes and regains control. After operations are back to normal, the supervisor becomes an observer again. From Figure 9.5 we see that our adaptive fuzzy control scheme operated in exactly this way.

Example 9.2. In this example, we consider the Duffing forced-oscillation system:

$$\dot{x}_1 = x_2 \tag{9.45}$$

$$\dot{x}_2 = -0.1x_2 - x_1^3 + 12cos(t) + u(t) \tag{9.46}$$

If without control, that is, $u(t) \equiv 0$, the system is chaotic. The trajectory of the system with $u(t) \equiv 0$ is shown in the (x_1, x_2) phase plane in Figure 9.6 for the initial condition $x_1(0) = x_2(0) = 2$ and time period $t_0 = 0$ to $t_f = 60$. We now use the first-type direct adaptive fuzzy controller to control the system state x_1 to track the reference trajectory $y_m(t) = sin(t)$. In the phase plane, this reference trajectory is the unit circle: $y_m^2 + \dot{y}_m^2 = 1$. We choose $k_1 = 2, k_2 = 1, \gamma = 2$, $Q = diag(10, 10)$, $M_\theta = 30$, $b_L = 1$ and $f^U = 12 + |x_1|^3$. We use the six fuzzy sets shown in Figure 9.2 for x_1 and x_2, and assume that there are no fuzzy control rules. We simulated two cases: $M_x = 10$ and $M_x = 3$. We directly integrated the differential equations of the closed-loop system and the adaptive law with stepsize 0.02. The closed-loop trajectories for the $M_x = 10$ and $M_x = 3$ cases are shown in Figures 9.7 and 9.8, respectively, where the initial condition $x_1(0) = x_2(0) = 2$

152 Stable Direct Adaptive Fuzzy Control of Nonlinear Systems   Chap. 9

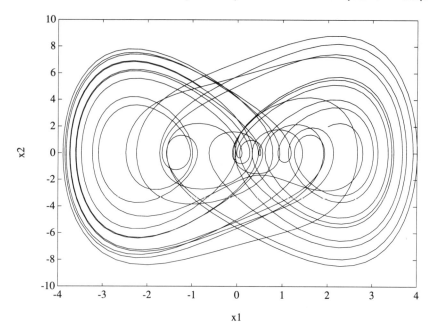

Figure 9.6 Trajectory of the chaotic system (9.45) and (9.46) in the (x_1, x_2) phase plane with $u(t) \equiv 0$ and $x_1(0) = x_2(0) = 2$.

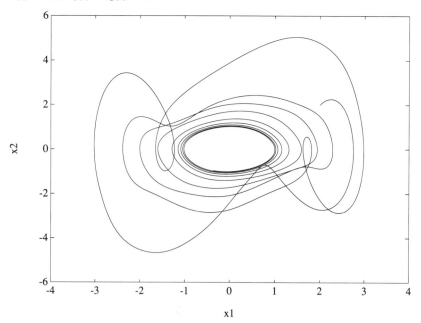

Figure 9.7 Closed-loop system trajectory $(x_1(t), x_2(t))$ using the first-type direct adaptive fuzzy controller for the chaotic system (9.45) and (9.46) for the $M_x = 10$ case.

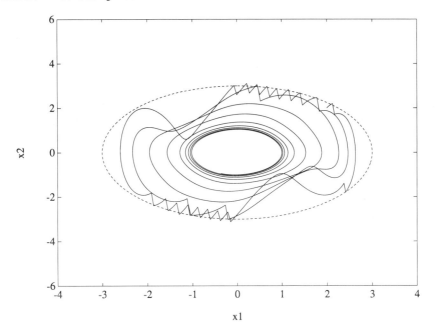

Figure 9.8 Closed-loop system trajectory $(x_1(t), x_2(t))$ using the first-type direct adaptive fuzzy controller for the chaotic system (9.45) and (9.46) for the $M_x = 3$ case.

and the trajectories were from $t_0 = 0$ to $t_f = 60$. We see that in both cases our adaptive fuzzy controller could control the system to track the reference trajectory. In Figure 9.7 the trajectory never hit the boundary $x_1^2 + x_2^2 = 10$ so that the supervisory control u_s never fired. Figure 9.8 shows how the supervisory control u_s forces the trajectory to be within the constraint set $x_1^2 + x_2^2 \leq 3$.

We also simulated the second-type direct adaptive fuzzy controller for these two examples, and the results are very similar to those in Figures 9.3–9.5, 9.7, and 9.8; we omit the details.

9.6 CONCLUDING REMARKS

In this chapter, we developed the first and second types of direct adaptive fuzzy controllers which: (1) do not require an accurate mathematical model of the system under control, (2) are capable of incorporating fuzzy control rules directly into the controllers, and (3) guarantee the global stability of the resulting closed-loop system in the sense that all signals involved are uniformly bounded. We provided the specific formula of the bounds so that controller designers can determine the bounds based on their requirements. We used the first-type direct adaptive fuzzy controller to regulate a first-order unstable system to the origin and to control the Duffing chaotic system to track a trajectory, and the simulation results showed that (1) the adaptive fuzzy controller could perform successful control without using any linguistic information, and (2) after incorporating some fuzzy control rules into the controllers, the adaptation speed became faster. We also showed explicitly how the supervisory control forced the state to be within the constraint set and how the adaptive fuzzy controller learned to regain control.

The indirect adaptive fuzzy controllers developed in the last chapter can incorporate fuzzy descriptions but cannot incorporate fuzzy control rules; on the other hand, the direct adaptive fuzzy controllers developed in this chapter can incorporate fuzzy control rules but cannot incorporate fuzzy descriptions. Thus, a natural question is: Is it possible to develop an adaptive fuzzy controller which can directly incorporate both fuzzy control rules and fuzzy descriptions? The answer is yes. In fact, this kind of adaptive fuzzy controller can be developed by using the same ideas as in the last and the present chapters, as follows.

Consider the system (9.1). Let u_i be an indirect adaptive fuzzy controller, that is,

$$u_i = \frac{1}{\hat{b}}(-\hat{f} + y_m^{(n)} + \underline{k}^T \underline{e}) \qquad (9.47)$$

where $\hat{b}$ is an estimate of b ($\hat{b}$ can be viewed as a fuzzy logic system with only one rule), $\hat{f}$ is a fuzzy logic system in the form of either (8.1) or (8.3), and $\underline{k}$ and $\underline{e}$ are defined as usual. Let u_d be a direct adaptive fuzzy controller, that is, u_d is a fuzzy logic system in the form of either (8.1) or (8.3). Let u_s be the usual supervisory control. We construct the overall control as follows:

$$u = \alpha u_i + (1 - \alpha)u_d + u_s \qquad (9.48)$$

where $\alpha \in [0, 1]$ is a weighting factor. If fuzzy control rules are more important and reliable than fuzzy descriptions, choose α to be small; otherwise, choose α to be large. Substituting (9.48) into (9.1), we have

$$x^{(n)} = \alpha(f + bu_i) + (1 - \alpha)(f + bu_d) + bu_s \qquad (9.49)$$

Let u^* be defined as in (9.4). Using the same kind of manipulation as in Sections 8.2 and 9.2, we obtain the following error equation governing the closed-loop system:

$$e^{(n)} = -\underline{k}^T \underline{e} + \alpha(\hat{f} - f) + \alpha(\hat{b} - b)u_i + (1 - \alpha)b(u^* - u_d) + bu_s \qquad (9.50)$$

Then using the same ideas as in Sections 8.2 and 9.2, we can specify the supervisory control u_s to bound the state and develop adaptive laws to adjust the parameter $\hat{b}$ and the parameters in $\hat{f}$ and u_d. Clearly, the control (9.48) can incorporate both fuzzy descriptions (due to u_i) and fuzzy control rules (due to u_d).

10

DESIGN OF ADAPTIVE FUZZY CONTROLLERS USING INPUT-OUTPUT LINEARIZATION CONCEPT

10.1 INTRODUCTION

The adaptive fuzzy controllers in Chapters 8 and 9 were designed for nonlinear systems in the canonical form (8.5) or (9.1). In practice, however, many nonlinear systems may not be represented in the canonical form. In general, single-input–single-output continuous-time nonlinear systems are described by

$$\dot{\underline{x}} = F(\underline{x}, u), \quad y = h(\underline{x}) \tag{10.1}$$

where $\underline{x} \in R^n$ is the state vector, $u \in R$ and $y \in R$ are the input and output of the system, respectively, and F and h are nonlinear functions. In this chapter, we consider the general nonlinear system (10.1); our objective is to make the output $y(t)$ track a desired trajectory $y_m(t)$.

Comparing (10.1) with the canonical form (8.5), we see that a difficulty with this general model is that the output y is only indirectly related to the input u, through the state variable $\underline{x}$ and the nonlinear state equation; on the other hand, with the canonical form the output is directly related to the input. Therefore, inspired by the results of Chapters 8 and 9, we might guess that the difficulty of the tracking control for the general nonlinear system (10.1) can be reduced if we can find *a direct relation between the system output y and the control input u*. Indeed, this idea constitutes the intuitive basis for the so-called *input-output linearization* approach to nonlinear control design [Slotine, Li, 66].

Input-output linearization is an approach to nonlinear control design which has attracted a great deal of interest in the nonlinear control community in recent years. This approach differs entirely from conventional linearization (for example,

Jacobian linearization) in that linearization is achieved by exact state transformations and feedback rather than by linear approximations of the nonlinear dynamics. We now briefly describe the basic concepts of linearization (Section 10.2), show how to design adaptive fuzzy controllers for the general nonlinear system (10.1) based on the input-output linearization concept (Section 10.3), and apply these adaptive fuzzy controllers to the ball-and-beam system (Section 10.4).

10.2 INTUITIVE CONCEPTS OF INPUT-OUTPUT LINEARIZATION

The basic idea of input-output linearization can be summarized as follows [Hauser, Sastry, Kokotovic, 18]: *differentiate the output y repeatedly until the input u appears, then design u to cancel the nonlinearity, and finally formulate a controller based on linear control.* We now specify this basic idea through an example which is taken from Slotine and Lee [66].

Consider the third-order system

$$\dot{x}_1 = sin(x_2) + (x_2 + 1)x_3 \tag{10.2}$$

$$\dot{x}_2 = x_1^5 + x_3 \tag{10.3}$$

$$\dot{x}_3 = x_1^2 + u \tag{10.4}$$

$$y = x_1 \tag{10.5}$$

To generate a direct relation between y and u, we differentiate y

$$\dot{y} = \dot{x}_1 = sin(x_2) + (x_2 + 1)x_3 \tag{10.6}$$

Since $\dot{y}$ is still not directly related to u, we differentiate it again

$$y^{(2)} = (cosx_2 + x_3)(x_1^5 + x_3) + (x_2 + 1)x_1^2 + (x_2 + 1)u \tag{10.7}$$

Clearly, (10.7) represents a direct relationship between y and u. If we choose

$$u = \frac{1}{x_2 + 1}(v - f_1) \tag{10.8}$$

where $f_1 \equiv (cosx_2 + x_3)(x_1^5 + x_3) + (x_2 + 1)x_1^2$, then we obtain

$$y^{(2)} = v \tag{10.9}$$

If we view v as a new input, then the original nonlinear system (10.1) is *linearized* to the linear system (10.9). This linearization procedure is shown in Figure 10.1. Now if we choose the new control input

$$v = y_m^{(2)} + k_1\dot{e} + k_2 e \tag{10.10}$$

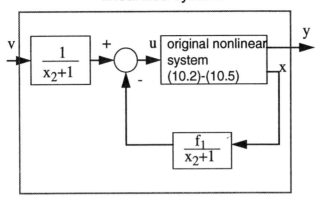

Figure 10.1 Diagram of linearized system of the nonlinear system (10.2)–(10.5).

where $e \equiv y_m - y$, then the closed-loop system is characterized by

$$e^{(2)} + k_1 \dot{e} + k_2 e = 0 \tag{10.11}$$

If we choose k_1 and k_2 such that all roots of $s^2 + k_1 s + k_2 = 0$ are in the open left half-plane, then we have $lim_{t \to \infty} e(t) = 0$—our control objective.

If we need to differentiate the output of a system r times to generate a direct relationship between the output y and the input u, this system is said to have *relative degree r*. Thus, the system (10.2)–(10.5) has relative degree 2. It can be shown formally that for any controllable system of order n, it will take at most n differentiations of any output for the control input to appear; that is, the relative degree of any nth-order controllable system is less than or equal to n.

At this point one might feel that the tracking control problem for the nonlinear system (10.2)–(10.5) has been solved with control law (10.8) and (10.10). However, one must realize that (10.11) only accounts for part of the whole system because it has only order 2, while the whole system has order 3. Therefore, a part of the system dynamics has been rendered "unobservable" in the input-output linearization. This part of the system is called the *internal dynamics* because it cannot be seen from the external input-output relationship (10.9). For the preceding example, the internal dynamics are represented by

$$\dot{x}_3 = x_1^2 + \frac{1}{x_2 + 1}(y_m^{(2)} + k_1 \dot{e} + k_2 e - f_1) \tag{10.12}$$

which is obtained by substituting (10.8) and (10.10) into (10.4). If these internal dynamics are stable in the sense that the internal state x_3 is bounded, our tracking control problem has indeed been solved. Otherwise, we have to redesign the control law.

10.3 DESIGN OF ADAPTIVE FUZZY CONTROLLERS BASED ON INPUT-OUTPUT LINEARIZATION

From Section 10.2 we see that the input-output linearization approach requires that the mathematical model of the system is given because otherwise the differentiation procedure cannot be performed. In our problem, however, the nonlinear functions F and h in (10.1) are assumed to be unknown. How can we generalize the concept of input-output linearization to this case?

First, inspired by the certainty equivalent controller in Section 8.2, one may think to replace the F and h by fuzzy logic systems, and to develop an adaptive law to adjust the parameters of the fuzzy logic systems such that they will approximate F and h. Indeed, this is the basic idea of Sastry and Isidori [64] where the F and h are approximated by series expansions of known nonlinear basis functions. However, as we see from Sastry and Isidori [64], this approach results in a very complicated adaptive control system. The reason is that although the original series expansions for approximating F and h are linear in their parameters, the differentiations cause these parameters to appear in a nonlinear fashion in later stages of the differentiation procedure. Therefore, we do not take this approach.

Our approach is based on the following consideration: from Section 10.2 we see that the control (10.8) is designed based only on the final system (10.7) in the differentiation procedure, the intermediate system (10.6) is not directly used; therefore, instead of approximating the F and h by fuzzy logic systems (or other series expansions of nonlinear basis functions as in Sastry and Isidori [64]), we approximate the nonlinear functions in the final equation of the differentiation procedure by fuzzy logic systems. Then we develop an adaptive law to adjust the parameters of the fuzzy logic systems for the purpose of making y to track y_m. To show the details in a formal way, we need the following assumption.

Assumption 10.1. We assume that: (1) the nonlinear system (10.1) has relative degree r; (2) the control u appears linearly with respect to $y^{(r)}$, that is,

$$y^{(r)} = f(\underline{x}) + g(\underline{x})u \qquad (10.13)$$

where f and g are unknown functions, and $g(\underline{x}) \neq 0$ for $\underline{x}$ in some controllable region U_c; and (3) the internal dynamics of the system with the following adaptive fuzzy controller are stable.

Design of an Adaptive Fuzzy Controller Based on Input-Output Linearization

Step 1:

Determine the relative degree r of the nonlinear system (10.1) based on physical intuitions. Specifically, we analyze the physical meanings of $y, \dot{y}, y^{(2)}, \ldots$, and determine the $y^{(r)}$ that is directly related to u. We show how to do this for the ball-and-beam system (Figure 4.2) in Section 10.4.

Sec. 10.3 Design of Fuzzy Controllers Based on Input-Output Linearization

Step 2:

Specify fuzzy logic systems $\hat{f}(\underline{x}|\underline{\theta}_f)$ and $\hat{g}(\underline{x}|\underline{\theta}_g)$ in the form of (8.1), that is,

$$\hat{f}(\underline{x}|\underline{\theta}_f) = \underline{\theta}_f^T \underline{\xi}(\underline{x}) \tag{10.14}$$

$$\hat{g}(\underline{x}|\underline{\theta}_g) = \underline{\theta}_g^T \underline{\xi}(\underline{x}) \tag{10.15}$$

using exactly the same procedure as in Step 2 of Section 8.3.

Step 3:

Choose the control

$$u = \frac{1}{\hat{g}(\underline{x}|\underline{\theta}_g)}[-\hat{f}(\underline{x}|\underline{\theta}_f) + y_m^{(r)} + \underline{k}^T \underline{e}] \tag{10.16}$$

where $\underline{k} = (k_r, \ldots, k_1)^T$ is such that all roots of $s^r + k_1 s^{r-1} + \cdots + k_r = 0$ are in the open left half-plane, and $\underline{e} = (e, \dot{e}, \ldots, e^{(r-1)})^T$ with $e = y_m - y$; and use the adaptive law

$$\underline{\dot{\theta}}_f = -\gamma_1 \underline{e}^T P \underline{b}_c \underline{\xi}(\underline{x}) \tag{10.17}$$

$$\underline{\dot{\theta}}_g = -\gamma_2 \underline{e}^T P \underline{b}_c \underline{\xi}(\underline{x}) u \tag{10.18}$$

to adjust the parameters, where P and $\underline{b}_c$ are defined as in (8.11) and (8.12). The overall control system is the same as Figure 8.1 except that the plant is changed to (10.1) and there is no supervisory control.

Remark 10.1. The control (10.16) and adaptive law (10.17) and (10.18) are obtained by using the same Lyapunov synthesis approach in Section 8.2. This can be done because from Assumption 10.1 we see that controlling (10.1) is equivalent to controlling (10.13) which is in the same form of (8.5). Therefore, the second and third parts of Theorem 8.1 apply to this adaptive fuzzy controller.

Remark 10.2. Using the same ideas as in Section 8.2, we can append a supervisory control to the fuzzy controller (10.16) to guarantee the boundedness of $\underline{y} = (y, \dot{y}, \ldots, y^{(r-1)})^T$. We can also use the projection algorithm to modify the adaptive law (10.17) and (10.18) to guarantee the boundedness of $\underline{\theta}_f$ and $\underline{\theta}_g$. If linguistic information on f and g is available, we can incorporate it into initial $\underline{\theta}_f(0)$ and $\underline{\theta}_g(0)$ in the same way as in Section 8.3.

Remark 10.3. The preceding adaptive fuzzy controller is a first-type indirect adaptive fuzzy controller. We can also develop a first-type direct adaptive fuzzy controller using the same idea as in Section 9.2. The resulting control is

$$u = \underline{\theta}^T \underline{\xi}(\underline{x}) \tag{10.19}$$

with adaptive law

$$\dot{\underline{\theta}} = \gamma \underline{e}^T \underline{p}_r \underline{\xi}(\underline{x}) \quad (10.20)$$

where $\underline{p}_r$ is defined as in Section 9.2. Similarly, we can also develop second-type adaptive fuzzy controllers.

10.4 APPLICATION TO THE BALL-AND-BEAM SYSTEM CONTROL

The ball-and-beam system is illustrated in Figure 4.2 and is characterized by (4.19) and (4.20). In this section, we use the adaptive fuzzy controllers developed in Section 10.3 to control the ball position $y = x_1$ to track the trajectory $y_m(t) = sin(t)$.

We begin with Step 1, that is, determining the relative degree r of the ball-and-beam system based on physical intuitions. First, we realize that the control u equals the acceleration of the beam angle θ; thus, our goal is to determine which derivative of y is directly related to the acceleration of θ. Clearly, the ball position $y = x_1 = r$ and the ball speed along the beam $\dot{y}$ are not directly related to u. Based on Newton's Law, the acceleration of the ball position, $y^{(2)}$, is propositional to $sin(\theta)$ which is not directly related to $u = \theta^{(2)}$. Therefore, $y^{(2)}$ is not directly related to u. Because $y^{(2)}$ is propositional to $sin(\theta)$, $y^{(3)}$ is directly related to $\dot{\theta}$ but not directly related to $u = \theta^{(2)}$. Finally, we see that $y^{(4)}$ is directly related to u. Therefore, the relative degree of the ball-and-beam system equals 4.[1]

In Step 2, we defined three fuzzy sets $\mu_N(x_i) = 1/(1 + exp(5(x_i + 1)))$, $\mu_Z(x_i) = exp(-x_i^2)$, and $\mu_P(x_i) = 1/(1 + exp(-5(x_i - 1)))$ for all x_1 to x_4. Therefore, the dimension of $\underline{\theta}_f, \underline{\theta}_g$ and $\underline{\xi}(\underline{x})$ equals $3^4 = 81$. Because there is no linguistic information about f and g, the initial $\underline{\theta}_f(0)$ and $\underline{\theta}_g(0)$ were chosen randomly in the interval $[-2, 2]$.

In Step 3, we chose $\underline{k} = (1, 4, 6, 4)^T$, $\gamma_1 = 2$, and $\gamma_2 = 0.2$. Figures 10.2 and 10.3 show the $y(t)$ (solid line) using this adaptive fuzzy controller along with the desired trajectory $y_m(t)$ (dashed line) for initial conditions $\underline{x}(0) = (1, 0, 0, 0)^T$ and $\underline{x}(0) = (-1, 0, 0, 0)^T$, respectively.

We also simulated the direct adaptive fuzzy controller (10.19) and (10.20). We chose the $\underline{\xi}(\underline{x})$ to be the same as in the preceding indirect adaptive fuzzy controller. We chose $\gamma = 10$. Figures 10.4 and 10.5 show the $y(t)$ (solid line) using this direct adaptive fuzzy controller along with the desired trajectory $y_m(t)$ for initial conditions $\underline{x}(0) = (1, 0, 0, 0)^T$ and $\underline{x}(0) = (-0.4, 0, 0, 0)^T$, respectively. Comparing Figures 10.2–10.3 with Figures 10.4–10.5, we see that the direct adaptive fuzzy controller gave better performance than the indirect adaptive fuzzy controller for this example.

---

[1] We did not give a mathematically rigorous definition of relative degree in Section 10.2; therefore, the relative degree determined in the preceding fashion can only be viewed as an intuitive relative degree. For the rigorous definition of relative degree, see Isidori [21] and Slotine and Li [66].

Sec. 10.4 Application to the Ball-and-Beam System Control   161

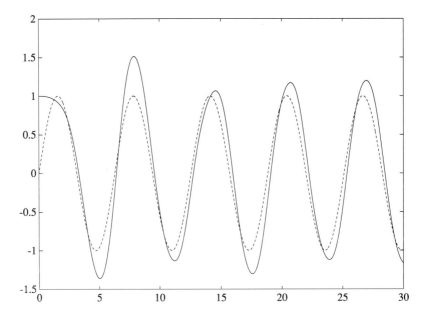

Figure 10.2 The output $y(t)$ (solid line) using the indirect adaptive fuzzy controller (10.16)–(10.18) and the desired trajectory $y_m(t)$ (dashed line) with the initial condition $\underline{x}(0) = (1, 0, 0, 0)^T$.

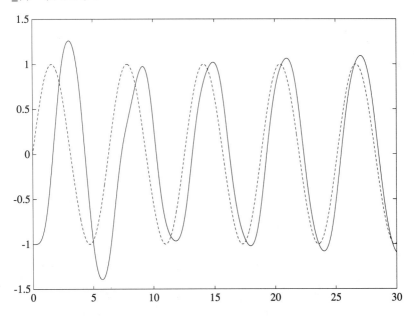

Figure 10.3 The output $y(t)$ (solid line) using the indirect adaptive fuzzy controller (10.16)–(10.18) and the desired trajectory $y_m(t)$ (dashed line) with the initial condition $\underline{x}(0) = (-1, 0, 0, 0)^T$.

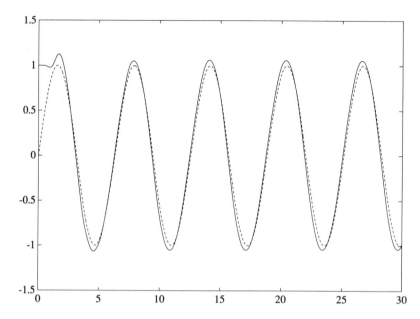

Figure 10.4 The output $y(t)$ (solid line) using the direct adaptive fuzzy controller (10.19)–(10.20) and the desired trajectory $y_m(t)$ (dashed line) with the initial condition $\underline{x}(0) = (1, 0, 0, 0)^T$.

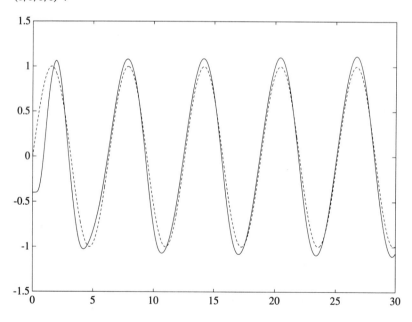

Figure 10.5 The output $y(t)$ (solid line) using the direct adaptive fuzzy controller (10.19)–(10.20) and the desired trajectory $y_m(t)$ (dashed line) with the initial condition $\underline{x}(0) = (-0.4, 0, 0, 0)^T$.

10.5 CONCLUDING REMARKS

In this chapter we showed the basic ideas of how to use the input-output linearization concept to construct adaptive fuzzy controllers for general nonlinear systems. We only gave the design steps of the adaptive fuzzy controllers and did not analyze their performance in a rigorous fashion as in Chapters 8 and 9. If Assumption 10.1 is true, the adaptive fuzzy controllers are well justified. However, the conditions in Assumption 10.1 are usually difficult to verify for practical problems. Therefore, we feel that this chapter is only a first step in the development of a comprehensive theory of adaptive fuzzy control for general nonlinear systems.

11

DESIGN AND STABILITY ANALYSIS OF FUZZY IDENTIFIERS OF NONLINEAR DYNAMIC SYSTEMS

11.1 INTRODUCTION

Existing identification schemes determine a model for a system based on the input-output pairs resulting from exciting the system with an input signal and measuring the corresponding outputs [Goodwin and Payne, 16; Ljung, 37]. For many industrial systems, there is another important information source—human operators, who are familiar with the systems and can provide linguistic descriptions about the behavior of the systems in terms of vague and fuzzy words. Although these linguistic descriptions are not precise, they provide important information about the system. In fact, for many industrial process control problems, a human operator can determine a set of successful control rules based only on the linguistic descriptions about the processes. Unfortunately, existing identification schemes ignore this important source of information and cannot incorporate the linguistic descriptions directly into the identifiers. The aim of this chapter is to develop identifiers of nonlinear dynamic systems which combine both linguistic descriptions and input-output pairs in a uniform fashion into their designs.

In Section 11.2, we show that fuzzy identifiers, which are identification models with adaptive fuzzy systems as their basic building blocks, are capable of following the outputs of any continuous-time nonlinear dynamic system to arbitrary accuracy. In Sections 11.3 and 11.4, we develop a first-type and a second-type fuzzy identifiers, respectively, and analyze their properties. As in the classification of adaptive fuzzy controllers, the first-type fuzzy identifiers use fuzzy logic systems in the form of (8.1), while the second-type fuzzy identifiers use fuzzy logic systems in the form of (8.3). In Section 11.5, we simulate the fuzzy identifiers to identify the chaotic glycolytic oscillator.

11.2 FUZZY IDENTIFIERS AS UNIVERSAL APPROXIMATORS OF DYNAMIC SYSTEMS[1]

Consider the dynamic system

$$\dot{\underline{x}} = A\underline{x} + g(\underline{x}, \underline{u}) \qquad (11.1)$$

$$\underline{y} = h(\underline{x}, \underline{u}) \qquad (11.2)$$

where $\underline{u} \in U \subset R^n$ is the input, $\underline{y} \in V \subset R^m$ is the output, $\underline{x} \in W \subset R^r$ is the state, A is a known Hurwitz matrix, and g, h are unknown continuous functions. Because g is unknown, (11.1) has no difference with $\dot{\underline{x}} = F(\underline{x}, \underline{u})$, where F is unknown; therefore, (11.1) and (11.2) represent a general continuous-time nonlinear dynamic system.

Our fuzzy identifier is constructed by replacing the g and h in (11.1) and (11.2) with fuzzy logic systems; that is,

$$\dot{\hat{\underline{x}}} = A\hat{\underline{x}} + \hat{g}(\hat{\underline{x}}, \underline{u}|\Theta_g) \qquad (11.3)$$

$$\hat{y} = \hat{h}(\hat{\underline{x}}, \underline{u}|\Theta_h) \qquad (11.4)$$

where $\hat{g}$ and $\hat{h}$ are fuzzy logic systems in the form of (2.46). The basic question is whether the output of the fuzzy identifier, $\hat{y}$, can follow the output of the system $\underline{y}$ to arbitrary accuracy. We now show that the answer is yes.

In order to give a formal analysis, we need the following assumption.

Assumption 11.1. For any $\underline{u}$ in compact U and any finite initial condition $\underline{x}(0) = \underline{x}^0$, the solution $\underline{x}(t)$ of (11.1) satisfies $|\underline{x}(t) - \underline{x}^0| \leq b$ for some positive constant b and all $t \in [0, T]$, where T is an arbitrary positive constant.

This assumption essentially requires that the system under identification is stable in the bounded-input–bounded-output sense. This is a reasonable assumption because we are rarely asked to identify an unstable system.

Define K to be the set

$$K = [(\underline{x}, \underline{u}) \in R^{r+n} : |\underline{x} - \underline{x}^0| \leq b + \epsilon, \underline{u} \in U] \qquad (11.5)$$

where $\epsilon > 0$ is an arbitrary constant, and $\underline{x}^0$ is any finite initial condition. Because U is compact and $|\underline{x}| \leq |\underline{x}^0| + b + \epsilon < \infty$, K is compact. Assumption 11.1 assures that for $\underline{u} \in U$ and finite $\underline{x}^0$, (11.1) generates $(\underline{x}(t), \underline{u}(t)) \in K$ for $t \in [0, T]$. Let $\hat{g}(\underline{x}, \underline{u}|\Theta_g) = (\hat{g}_1(\underline{x}, \underline{u}|\Theta_g), \ldots, \hat{g}_r(\underline{x}, \underline{u}|\Theta_g))^T$ and $\hat{h}(\underline{x}, \underline{u}|\Theta_h) = (\hat{h}_1(\underline{x}, \underline{u}|\Theta_h), \ldots, \hat{h}_m(\underline{x}, \underline{u}|\Theta_h))^T$ be the fuzzy logic systems in the form of (2.46), where Θ_g, Θ_h are collections of the parameters $a_i^l, \bar{x}_i^l, \sigma_i^l$ and $\bar{y}^l$. Define Θ_g^* and Θ_h^* be such that

$$sup_{\underline{x},\underline{u}\in K}|\hat{g}(\underline{x}, \underline{u}|\Theta_g^*) - g(\underline{x}, \underline{u})| < \epsilon_g \qquad (11.6)$$

$$sup_{\underline{x},\underline{u}\in K}|\hat{h}(\underline{x}, \underline{u}|\Theta_h^*) - h(\underline{x}, \underline{u})| < \epsilon_h \qquad (11.7)$$

---

[1] The results in the section were inspired by Polycarpou and Ioannou [57]

for some arbitrary $\epsilon_g > 0$ and $\epsilon_h > 0$. The Universal Approximation Theorem in Chapter 2 assures the existence of the Θ_g^* and Θ_h^*. We now show that by replacing g and h in (11.1) and (11.2) with the $\hat{g}$ and $\hat{h}$, we obtain a dynamic system whose output can approximate the output of (11.1) and (11.2) to arbitrary accuracy over any finite interval of time.

Theorem 11.1. Consider the identification model

$$\dot{\hat{x}} = A\hat{x} + \hat{g}(\hat{x}, \underline{u}|\Theta_g^*) \tag{11.8}$$

$$\hat{y} = \hat{h}(\hat{x}, \underline{u}|\Theta_h^*) \tag{11.9}$$

where $\hat{x}(0) = \underline{x}(0) = \underline{x}^0$, $\underline{u} \in U$, $\hat{g}$ and $\hat{h}$ are the fuzzy logic systems in the form of (2.46), and the parameters Θ_g^* and Θ_h^* are defined in (11.6) and (11.7). Then, for any $\epsilon > 0$, finite $T > 0$, and properly chosen ϵ_g and ϵ_h, we have that

$$sup_{t\in[0,T]}|\hat{y}(t) - \underline{y}(t)| < \epsilon \tag{11.10}$$

Proof of this theorem is given in the Appendix.

11.3 DESIGN AND STABILITY ANALYSIS OF FIRST-TYPE FUZZY IDENTIFIERS

Consider the identification of the nonlinear system

$$\dot{\underline{x}} = f(\underline{x}) + g(\underline{x})u \tag{11.11}$$

where f and g are unknown functions from $U = U_1 \times \cdots \times U_n$ to $V = V_1 \times \cdots \times V_n$, $U_i, V_i \subset R$, $i = 1, 2, \ldots, n$, and the input $u \in R$ and the state $\underline{x} \in R^n$ are assumed to be bounded and available for measurement. Our purpose is to develop an identification model where the f and g are replaced by fuzzy logic systems $\hat{f}(*|\Theta_f)$ and $\hat{g}(*|\Theta_g)$ and an adaptive law for updating the parameter matrices Θ_f and Θ_g such that the identification model converges to the real system (11.11).[2] First, we rewrite (11.11) as

$$\dot{\underline{x}} = \hat{f}(\underline{x}|\Theta_f^*) + \hat{g}(\underline{x}|\Theta_g^*)u + \underline{w} \tag{11.12}$$

where

$$\underline{w} \equiv [f(\underline{x}) - \hat{f}(\underline{x}|\Theta_f^*)] + [g(\underline{x}) - \hat{g}(\underline{x}|\Theta_g^*)]u \tag{11.13}$$

$$\Theta_f^* \equiv argmin_{\Theta_f \in \Omega_f}[sup_{\underline{x} \in U}|f(\underline{x}) - \hat{f}(\underline{x}|\Theta_f)|] \tag{11.14}$$

$$\Theta_g^* \equiv argmin_{\Theta_g \in \Omega_g}[sup_{\underline{x} \in U}|g(\underline{x}) - \hat{g}(\underline{x}|\Theta_g)|] \tag{11.15}$$

and Ω_f and Ω_g are bounded feasible sets of Θ_f and Θ_g, respectively. We assume that $\Omega_f \equiv [\Theta : tr(\Theta\Theta^T) \leq M_f]$ and $\Omega_g \equiv [\Theta : tr(\Theta\Theta^T) \leq M_g]$, where M_f and

---

[1] In the first-type fuzzy identifier we collect the parameters of the fuzzy logic systems into matrices; in the second-type fuzzy identifier we collect the parameters of the fuzzy logic systems into vectors.

Sec 11.3 Design and Stability Analysis of First-Type Fuzzy Identifiers

M_g are given constants. Because of the Universal Approximation Theorem and the boundedness of u, $|\underline{w}|$ can be arbitrarily small; therefore, without loss of generality we assume that $w_0 \equiv sup_{t \geq 0}|\underline{w}(t)|$ is finite. We use the following series-parallel identification model (see Chapter 3):

$$\dot{\hat{\underline{x}}} = -\alpha \hat{\underline{x}} + \alpha \underline{x} + \hat{f}(\underline{x}|\Theta_f) + \hat{g}(\underline{x}|\Theta_g)u \qquad (11.16)$$

where α is a given positive scalar. The whole identification scheme is shown in Figure 11.1. The goals of identification are:

Identification Goals. Specify the fuzzy logic systems $\hat{f}(\underline{x}|\Theta_f)$ and $\hat{g}(\underline{x}|\Theta_g)$, and develop an adaptive law for the parameters Θ_f and Θ_g such that:

1. All signals involved in the identification model must be uniformly bounded. That is, it must be guaranteed that $\hat{\underline{x}} \in L_\infty$, $tr(\Theta_f \Theta_f^T) \leq M_f$, and $tr(\Theta_g \Theta_g^T) \leq M_g$ (the input u and the system state $\underline{x}$ are uniformly bounded by assumption).
2. The error $\underline{e} \equiv \underline{x} - \hat{\underline{x}}$ should be as small as possible.

To emphasize the point that our fuzzy identifiers can make use of linguistic descriptions about the unknown system (11.11), we make the following assumption.

Assumption 11.2. There are the following linguistic descriptions about the unknown functions $f(\underline{x})$ and $g(\underline{x})$:

$$R_f^r : \text{IF } x_1 \text{ is } A_1^r \text{ and } \cdots \text{ and } x_n \text{ is } A_n^r,$$
$$\text{THEN } f_1(\underline{x}) \text{ is } C_1^r, \cdots, f_n(\underline{x}) \text{ is } C_n^r \qquad (11.17)$$

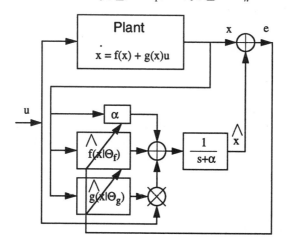

Initial $\Theta_f(0)$, $\Theta_g(0)$ are determined based on the linguistic descriptions about f(x) and g(x).

Figure 11.1 Identification model using adaptive fuzzy systems.

and

$$R_g^s : \text{IF } x_1 \text{ is } B_1^s \text{ and } \cdots \text{ and } x_n \text{ is } B_n^s,$$
$$\text{THEN } g_1(\underline{x}) \text{ is } D_1^s, \cdots, g_n(\underline{x}) \text{ is } D_n^s \qquad (11.18)$$

respectively, where A_i^r and B_i^s are fuzzy sets in U_i, C_i^r and D_i^s are fuzzy sets in V_i which achieve membership value 1 at some point, $r = 1, 2, \ldots, N_f$, and $s = 1, 2, \ldots, N_g$. We allow $N_f = N_g = 0$ which means that there are no linguistic descriptions about f and g.

Design of First-Type Fuzzy Identifiers

Step 1:

Define m_i fuzzy sets F_i^{ji} in U_i such that for any $x_i \in U_i$, there exists at least one $\mu_{F_i^{ji}}(x_i) \neq 0$, where $i = 1, 2, \ldots, n$, and $ji = 1, 2, \ldots, m_i$ (note that ji is a single index). We require that these F_i^{ji}'s include the A_i^r's and B_i^s's in (11.17) and (11.18). These fuzzy sets are fixed and will not change during the adaptation procedure of Step 4.

Step 2:

Construct the fuzzy rule base of fuzzy logic system $\hat{f}(\underline{x}|\Theta_f)$ which consists of the following $\prod_{i=1}^{n} m_i$ rules

$$R^l : \text{IF } x_1 \text{ is } F_1^{j1} \text{ and } \cdots \text{ and } x_n \text{ is } F_n^{jn},$$
$$\text{THEN } \hat{f}_1(\underline{x}) \text{ is } G_1^l, \cdots, \hat{f}_n(\underline{x}) \text{ is } G_n^l \qquad (11.19)$$

where F_i^{ji} are defined in Step 1, G_i^l are fuzzy sets in V_i with $\mu_{G_i^l}(\theta_{if}^l) = 1$ for some parameter $\theta_{if}^l \in V_i$, $\Theta_f = (\underline{\theta}_{1f}, \cdots, \underline{\theta}_{nf})^T$, $\underline{\theta}_{if} = (\theta_{if}^1, \theta_{if}^2, \ldots, \theta_{if}^{\prod_{i=1}^{n} m_i})^T$, $ji = 1, 2, \ldots, m_i$, $i = 1, 2, \ldots, n$, and $l = 1, 2, \ldots, \prod_{i=1}^{n} m_i$ with each l corresponding to a combination of $(j1, \ldots, jn)$. Because the fuzzy rule base consists of all the possible rules concerning the fuzzy sets F_i^{ji} of Step 1 which include the A_i^r's, it includes the linguistic descriptions in (11.17).

The initial parameters $\theta_{if}^l(0)$ are chosen as follows: if the IF part in (11.19) agrees with an IF part in (11.17), choose $\theta_{if}^l(0) = argsup_{y_i \in V_i}[\mu_{C_i^r}(y_i)]$; otherwise, choose $\theta_{if}^l(0)$ arbitrarily in V_i. Therefore, *we incorporate the linguistic descriptions into the fuzzy identifier by constructing the initial fuzzy identifier based on these descriptions.*

The fuzzy rule base and initial parameters of $\hat{g}(\underline{x}|\Theta_g)$ are determined in exactly the same way as $\hat{f}(\underline{x}|\Theta_f)$; we omit the details.

Step 3:

Choose $\hat{f}(\underline{x}|\Theta_f) = (\hat{f}_1(\underline{x}|\underline{\theta}_{1f}), \ldots, \hat{f}_n(\underline{x}|\underline{\theta}_{nf}))^T$ and $\hat{g}(\underline{x}|\Theta_g) = (\hat{g}_1(\underline{x}|\underline{\theta}_{1g}), \ldots, \hat{g}_n(\underline{x}|\underline{\theta}_{ng}))^T$ to be fuzzy logic systems with singleton fuzzifier, center average defuzzifier, and product inference (Lemma 2.1) based on the fuzzy rule bases constructed in Step 2,

$$\hat{f}_i(\underline{x}|\underline{\theta}_{if}) = \frac{\sum_{j1=1}^{m_1} \cdots \sum_{jn=1}^{m_n} \theta_{if}^l \left[\mu_{F_1^{j1}}(x_1) \cdots \mu_{F_n^{jn}}(x_n) \right]}{\sum_{j1=1}^{m_1} \cdots \sum_{jn=1}^{m_n} \left[\mu_{F_1^{j1}}(x_1) \cdots \mu_{F_n^{jn}}(x_n) \right]} \qquad (11.20)$$

Sec 11.3 Design and Stability Analysis of First-Type Fuzzy Identifiers

$$\hat{g}_i(\underline{x}|\underline{\theta}_{ig}) = \frac{\sum_{j1=1}^{m_1} \cdots \sum_{jn=1}^{m_n} \theta_{ig}^l \left[\mu_{F_1^{j1}}(x_1) \cdots \mu_{F_n^{jn}}(x_n) \right]}{\sum_{j1=1}^{m_1} \cdots \sum_{jn=1}^{m_n} \left[\mu_{F_1^{j1}}(x_1) \cdots \mu_{F_n^{jn}}(x_n) \right]} \quad (11.21)$$

where $i = 1, 2, \ldots, n$, and $l = 1, 2, \ldots, \prod_{i=1}^{n} m_i$ with each l corresponding to a combination of $(j1, \ldots, jn)$. Defining the fuzzy basis function

$$\xi^l(\underline{x}) = \frac{\mu_{F_1^{j1}}(x_1) \cdots \mu_{F_n^{jn}}(x_n)}{\sum_{j1=1}^{m_1} \cdots \sum_{jn=1}^{m_n} \left[\mu_{F_1^{j1}}(x_1) \cdots \mu_{F_n^{jn}}(x_n) \right]} \quad (11.22)$$

where l is defined as in (11.20) and (11.21). Collecting them into an $\prod_{i=1}^{n} m_i \times 1$ vector $\underline{\xi}(\underline{x})$, and collecting the θ_{if}^l and θ_{ig}^l into $\prod_{i=1}^{n} m_i \times 1$ vectors $\underline{\theta}_{if}$ and $\underline{\theta}_{ig}$ in the same order as $\underline{\xi}(\underline{x})$, we can rewrite (11.20) and (11.21) as

$$\hat{f}_i(\underline{x}|\underline{\theta}_{if}) = \underline{\theta}_{if}^T \underline{\xi}(\underline{x}) \quad (11.23)$$

and

$$\hat{g}_i(\underline{x}|\underline{\theta}_{ig}) = \underline{\theta}_{ig}^T \underline{\xi}(\underline{x}) \quad (11.24)$$

respectively. Recall that $\Theta_f \equiv (\underline{\theta}_{1f}, \ldots, \underline{\theta}_{nf})^T$ and $\Theta_g \equiv (\underline{\theta}_{1g}, \ldots, \underline{\theta}_{ng})^T$ and substituting (11.23) and (11.24) into (11.16), the fuzzy identifier becomes

$$\dot{\hat{\underline{x}}} = -\alpha \hat{\underline{x}} + \alpha \underline{x} + \Theta_f \underline{\xi}(\underline{x}) + \Theta_g \underline{\xi}(\underline{x}) u \quad (11.25)$$

Step 4:

Update the parameter matrices Θ_f and Θ_g using the following adaptive law:

$$\dot{\Theta}_f = \begin{cases} \gamma_1 \underline{e} \underline{\xi}^T(\underline{x}) & \text{if } (tr(\Theta_f \Theta_f^T) < M_f) \text{ or } (tr(\Theta_f \Theta_f^T) = M_f \text{ and } \underline{e}^T \Theta_f \underline{\xi}(\underline{x}) \leq 0) \\ P\left[\gamma_1 \underline{e} \underline{\xi}^T(\underline{x})\right] & \text{if } (tr(\Theta_f \Theta_f^T) = M_f \text{ and } \underline{e}^T \Theta_f \underline{\xi}(\underline{x}) > 0) \end{cases} \quad (11.26)$$

$$\dot{\Theta}_g = \begin{cases} \gamma_2 \underline{e} \underline{\xi}^T(\underline{x}) u & \text{if } (tr(\Theta_g \Theta_g^T) < M_g) \text{ or } (tr(\Theta_g \Theta_g^T) = M_g \text{ and } \underline{e}^T \Theta_g \underline{\xi}(\underline{x}) u \leq 0) \\ P\left[\gamma_2 \underline{e} \underline{\xi}^T(\underline{x}) u\right] & \text{if } (tr(\Theta_g \Theta_g^T) = M_g \text{ and } \underline{e}^T \Theta_g \underline{\xi}(\underline{x}) u > 0) \end{cases} \quad (11.27)$$

where γ_1 and γ_2 are positive constants, $P[*]$ is the projection operator defined as

$$P\left[\gamma_1 \underline{e} \underline{\xi}^T(\underline{x})\right] \equiv \gamma_1 \underline{e} \underline{\xi}^T(\underline{x}) - \gamma_1 \frac{\underline{e}^T \Theta_f \underline{\xi}(\underline{x})}{tr(\Theta_f \Theta_f^T)} \Theta_f \quad (11.28)$$

$$P\left[\gamma_2 \underline{e} \underline{\xi}^T(\underline{x}) u\right] \equiv \gamma_2 \underline{e} \underline{\xi}^T(\underline{x}) u - \gamma_2 \frac{\underline{e}^T \Theta_g \underline{\xi}(\underline{x}) u}{tr(\Theta_g \Theta_g^T)} \Theta_g \quad (11.29)$$

$\underline{e} = \underline{x} - \hat{\underline{x}}$, and the initial $\Theta_f(0)$ and $\Theta_g(0)$ are determined in Step 2.

Properties of the preceding fuzzy identifier are summarized in the following theorem.

Theorem 11.2. The fuzzy identifier (11.25) with the adaptive law (11.26) and (11.27) guarantees the following properties:

1. $tr(\Theta_f \Theta_f^T) \leq M_f$, $tr(\Theta_g \Theta_g^T) \leq M_g$, and $\hat{\underline{x}} \in L_\infty$.
2. There exist constants a and b such that

$$\int_0^t |\underline{e}(\tau)|^2 d\tau \leq a + b \int_0^t |\underline{w}(\tau)|^2 d\tau \qquad (11.30)$$

for any $t \geq 0$, where $\underline{w}(\tau)$ is defined by (11.13).
3. If $\underline{w} \in L_2$, then $lim_{t \to \infty} |\underline{e}(t)| = 0$.

Proof of this theorem is given in the Appendix.

Remark 11.1. From the third part of Theorem 11.2 we see that in order to have $lim_{t \to \infty} |\underline{e}(t)| = 0$, we require that $\underline{w}$ is squared integrable. From the definition of $\underline{w}$ (11.13)–(11.15), we see that $\underline{w}$ is the sum of the minimum approximation errors of $\hat{f}$ and $\hat{g}$ to f and g over the entire state space U. From the Universal Approximation Theorem, this minimum approximation error should be small if we properly choose the fuzzy logic systems $\hat{f}$ and $\hat{g}$.

Remark 11.2. From Step 2 we see that the linguistic descriptions (11.17) and (11.18) about the unknown functions $f(\underline{x})$ and $g(\underline{x})$ are used to construct the initial fuzzy identifier. If these descriptions provide good pictures of $f(\underline{x})$ and $g(\underline{x})$, we can expect that the adaptation procedure will converge fast because the initial identifier constructed from good descriptions should be close to the true system. If no linguistic descriptions are available, the fuzzy identifier becomes a regular nonlinear identifier, similar to the radial basis function [Chen, Cowan, and Grant, 5] and neural network [Narendra and Parthasarathy, 49] identifiers.

Remark 11.3. In this first-type fuzzy identifier, we fix the fuzzy sets in U and consider fuzzy logic systems whose IF parts are concerned only with these fuzzy sets. An advantage of doing so is that the fuzzy logic systems in the fuzzy identifier are linear in the parameter. Therefore, (1) we were able to use a relatively simpler adaptive law to update the parameters; and (2) convergence of the adaptation procedure is expected to be faster because we are not concerned with complicated nonlinear search problems. A disadvantage is that we have to consider all the possible combinations of the fuzzy sets in U because these fuzzy sets cannot change so that we should have rules to cover every region of U, where by *cover* we mean that for each $\underline{x} \in U$ there should exist at least one rule in the fuzzy rule bases of $\hat{f}$ and $\hat{g}$ whose strength $(\mu_{F_1^{j1}}(x_1) \cdots \mu_{F_n^{jn}}(x_n))$ is not very small. Therefore, if the dimension n of a problem is large, then we have to choose the m_i's relatively small so that the computational requirements of the fuzzy identifier do not exceed the capability of the computing sources available. Clearly, the bigger the m_i's, the larger the number of rules, and therefore the smaller the minimum approximation error $\underline{w}$. This in turn means the smaller the identification error $\underline{e}$ (see (11.30)). In practical applications of this fuzzy identifier, we have to find a compromise between complexity and accuracy.

11.4 DESIGN AND STABILITY ANALYSIS OF SECOND-TYPE FUZZY IDENTIFIERS

As discussed in Remark 11.3, the fuzzy identifier in Section 11.3 may require a large number of rules for higher dimensional systems. An obvious way to overcome this rule explosion problem is to allow the fuzzy sets in U also to change during the adaptation procedure so that in principle any rule can cover any region of U. Therefore, we only need a small number of rules. This is the basic idea of the second-type fuzzy identifier in this section. The price paid for this additional freedom is that the fuzzy identifier becomes nonlinear in the parameter, so that we have to use a more complicated adaptive law.

We consider the identification of the same system (11.11) and use the same series-parallel model (11.16) (Figure 11.1). The identification goals remain the same as in Section 11.3, and we still consider Assumption 11.2. However, the design of the second-type fuzzy identifier is different from the first one.

Design of Second-Type Fuzzy Identifiers

Step 1:

Define $2N$ fuzzy sets F_i^l and G_i^l in each U_i of U and $2N$ fuzzy sets E_i^l and H_i^l in each V_i of V with the following membership functions

$$\mu_{F_i^l}(x_i) = exp\left[-\frac{1}{2}\frac{(x_i - \bar{x}_{if}^l)^2}{(\sigma_{if}^l)^2 + \epsilon}\right] \quad (11.31)$$

$$\mu_{G_i^l}(x_i) = exp\left[-\frac{1}{2}\frac{(x_i - \bar{x}_{ig}^l)^2}{(\sigma_{ig}^l)^2 + \epsilon}\right] \quad (11.32)$$

$$\mu_{E_i^l}(y_i) = exp\left[-\frac{1}{2}\frac{(y_i - \bar{y}_{ie}^l)^2}{(\sigma_{ie}^l)^2 + \epsilon}\right] \quad (11.33)$$

$$\mu_{H_i^l}(y_i) = exp\left[-\frac{1}{2}\frac{(y_i - \bar{y}_{ih}^l)^2}{(\sigma_{ih}^l)^2 + \epsilon}\right] \quad (11.34)$$

where $\epsilon > 0$ is a small constant, $l = 1, 2, \ldots, N$ (in general, $N << \prod_{i=1}^n m_i$), $x_i \in U_i$, $y_i \in V_i$, $i = 1, 2, \ldots, n$, and $\bar{x}_{if}^l, \sigma_{if}^l, \bar{x}_{ig}^l, \sigma_{ig}^l, \bar{y}_{ie}^l, \sigma_{ie}^l, \bar{y}_{ih}^l$, and σ_{ih}^l are free parameters which will be updated in the adaptation procedure of Step 4. The purpose of adding the small $\epsilon > 0$ to the fuzzy membership functions (11.31)–(11.34) is that even if the σ's $= 0$, the fuzzy membership functions are still well defined. This modification will make the adaptive law simpler because we do not require the σ's $\neq 0$. For this fuzzy identifier, we assume that the membership functions of A_i^r, B_i^s, C_i^r, and D_i^s in (11.17) and (11.18) are in the form of (11.31)–(11.34), respectively, and that $N \geq N_f$ and $N \geq N_g$.

Step 2:

Construct the fuzzy rule bases of $\hat{f}$ and $\hat{g}$ as

$$R_f^l : \text{IF } x_1 \text{ is } F_1^l \text{ and } \cdots \text{ and } x_n \text{ is } F_n^l,$$
$$\text{THEN } \hat{f}_1(\underline{x}) \text{ is } E_1^l, \cdots, \hat{f}_n(\underline{x}) \text{ is } E_n^l \quad (11.35)$$

and

$$R_g^l : \text{IF } x_1 \text{ is } G_1^l \text{ and } \cdots \text{ and } x_n \text{ is } G_n^l,$$
$$\text{THEN } \hat{g}_1(\underline{x}) \text{ is } H_1^l, \cdots, \hat{g}_n(\underline{x}) \text{ is } H_n^l \quad (11.36)$$

respectively, where F_i^l, E_i^l, G_i^l, and H_i^l are defined in Step 1, and $l = 1, 2, \ldots, N$. Choose $\hat{f}(\underline{x}|\underline{\theta}_f) = (\hat{f}_1(\underline{x}|\underline{\theta}_f), \cdots, \hat{f}_n(\underline{x}|\underline{\theta}_f))^T$ and $\hat{g}(\underline{x}|\underline{\theta}_g) = (\hat{g}_1(\underline{x}|\underline{\theta}_g), \cdots, \hat{g}_n(\underline{x}|\underline{\theta}_g))^T$ to be fuzzy logic systems with singleton fuzzifier, modified center average defuzzifier, product inference, and Gaussian membership function (Lemma 2.4) based on the fuzzy rule bases of (11.35) and (11.36), respectively,

$$\hat{f}_i(\underline{x}|\underline{\theta}_f) = \frac{\sum_{l=1}^{N} \bar{y}_{ie}^l \left[\prod_{j=1}^{n} exp\left(-\frac{1}{2}\frac{(x_j - \bar{x}_{jf}^l)^2}{(\sigma_{jf}^l)^2 + \epsilon}\right) \right] / \left[(\sigma_{ie}^l)^2 + \epsilon\right]}{\sum_{l=1}^{N} \left[\prod_{j=1}^{n} exp\left(-\frac{1}{2}\frac{(x_j - \bar{x}_{jf}^l)^2}{(\sigma_{jf}^l)^2 + \epsilon}\right) \right] / \left[(\sigma_{ie}^l)^2 + \epsilon\right]} \quad (11.37)$$

$$\hat{g}_i(\underline{x}|\underline{\theta}_g) = \frac{\sum_{l=1}^{N} \bar{y}_{ih}^l \left[\prod_{j=1}^{n} exp\left(-\frac{1}{2}\frac{(x_j - \bar{x}_{jg}^l)^2}{(\sigma_{jg}^l)^2 + \epsilon}\right) \right] / \left[(\sigma_{ih}^l)^2 + \epsilon\right]}{\sum_{l=1}^{N} \left[\prod_{j=1}^{n} exp\left(-\frac{1}{2}\frac{(x_j - \bar{x}_{jg}^l)^2}{(\sigma_{jg}^l)^2 + \epsilon}\right) \right] / \left[(\sigma_{ih}^l)^2 + \epsilon\right]} \quad (11.38)$$

where $\underline{\theta}_f$ and $\underline{\theta}_g$ are $4nN \times 1$ vectors which are collections of the free parameters. Specifically, $\underline{\theta}_f \equiv (\bar{y}_{1e}^1, \ldots, \bar{y}_{ne}^1, \ldots, \bar{y}_{1e}^N, \ldots, \bar{y}_{ne}^N, \sigma_{1e}^1, \ldots, \sigma_{ne}^1, \ldots, \sigma_{1e}^N, \ldots, \sigma_{ne}^N, \bar{x}_{1f}^1, \ldots, \bar{x}_{nf}^1, \ldots, \bar{x}_{1f}^N, \ldots, \bar{x}_{nf}^N, \sigma_{1f}^1, \ldots, \sigma_{nf}^1, \ldots, \sigma_{1f}^N, \ldots, \sigma_{nf}^N)$, and $\underline{\theta}_g$ is defined in exactly the same way as $\underline{\theta}_f$ with the subscript e replaced by h and f replaced by g. Different from the first-type fuzzy identifier where we collected the free parameters into matrices Θ_f and Θ_g, we collect the free parameters into vectors $\underline{\theta}_f$ and $\underline{\theta}_g$ in this second-type fuzzy identifier. We still use the series-parallel identification model (11.16), with $\hat{f}(\underline{x}|\Theta_f)$ and $\hat{g}(\underline{x}|\Theta_g)$ replaced by the $\hat{f}(\underline{x}|\underline{\theta}_f)$ and $\hat{g}(\underline{x}|\underline{\theta}_g)$ of (11.37) and (11.38), respectively.

Step 3:

Compute the gradient matrices $\frac{\partial \hat{f}(\underline{x}|\underline{\theta}_f)}{\partial \underline{\theta}_f} = (\frac{\partial \hat{f}_1(\underline{x}|\underline{\theta}_f)}{\partial \underline{\theta}_f}, \ldots, \frac{\partial \hat{f}_n(\underline{x}|\underline{\theta}_f)}{\partial \underline{\theta}_f})$ and $\frac{\partial \hat{g}(\underline{x}|\underline{\theta}_g)}{\partial \underline{\theta}_g} = (\frac{\partial \hat{g}_1(\underline{x}|\underline{\theta}_g)}{\partial \underline{\theta}_g}, \ldots, \frac{\partial \hat{g}_n(\underline{x}|\underline{\theta}_g)}{\partial \underline{\theta}_g})$ using the following back-propagation algorithm (which is derived in the same manner as in Chapter 3):

$$\frac{\partial \hat{f}_i}{\partial \bar{y}_{ie}^l} = \frac{z^l}{b_i \left((\sigma_{ie}^l)^2 + \epsilon\right)} \quad (11.39)$$

$$\frac{\partial \hat{f}_i}{\partial \sigma_{ie}^l} = \frac{\bar{y}_{ie}^l - \hat{f}_i}{b_i} \frac{-2z^l \sigma_{ie}^l}{\left((\sigma_{ie}^l)^2 + \epsilon\right)^2} \quad (11.40)$$

Sec 11.4 Design and Stability Analysis of Second-Type Fuzzy Identifiers

$$\frac{\partial \hat{f}_i}{\partial \bar{x}_{jf}^l} = \frac{\bar{y}_{ie}^l - \hat{f}_i}{b_i} \frac{1}{\left(\sigma_{ie}^l\right)^2 + \epsilon} z^l \frac{x_j - \bar{x}_{jf}^l}{\left(\sigma_{jf}^l\right)^2 + \epsilon} \quad (11.41)$$

$$\frac{\partial \hat{f}_i}{\partial \sigma_{jf}^l} = \frac{\bar{y}_{ie}^l - \hat{f}_i}{b_i} \frac{1}{\left(\sigma_{ie}^l\right)^2 + \epsilon} z^l \frac{2\left(x_j - \bar{x}_{jf}^l\right)^2 \sigma_{jf}^l}{\left(\left(\sigma_{jf}^l\right)^2 + \epsilon\right)^2} \quad (11.42)$$

where

$$z^l \equiv \prod_{j=1}^{n} exp\left(-\frac{1}{2} \frac{\left(x_j - \bar{x}_{jf}^l\right)^2}{\left(\sigma_{jf}^l\right)^2 + \epsilon}\right) \quad (11.43)$$

$$b_i \equiv \sum_{l=1}^{n} z^l / \left(\left(\sigma_{ie}^l\right)^2 + \epsilon\right) \quad (11.44)$$

$l = 1, 2, \ldots, N$, and $i, j = 1, 2, \ldots, n$; and, $\frac{\partial \hat{g}_i}{\partial \bar{y}_{ih}^l}$, $\frac{\partial \hat{g}_i}{\partial \sigma_{ih}^l}$, $\frac{\partial \hat{g}_i}{\partial \bar{x}_{jg}^l}$, and $\frac{\partial \hat{g}_i}{\partial \sigma_{jg}^l}$ are computed in exactly the same way as (11.39)–(11.42), respectively, with the subscripts e replaced by h and f replaced by g.

Step 4:

Update the parameter vectors $\underline{\theta}_f$ and $\underline{\theta}_g$ using the following adaptive law:

$$\dot{\underline{\theta}}_f = \begin{cases} \gamma_1 \frac{\partial \hat{f}(\underline{x}|\underline{\theta}_f)}{\partial \underline{\theta}_f} e & \text{if } \left(|\underline{\theta}_f|^2 < M_f\right) \text{ or } \left(|\underline{\theta}_f|^2 = M_f \text{ and } \underline{e}^T \left(\frac{\partial \hat{f}}{\partial \underline{\theta}_f}\right)^T \underline{\theta}_f \leq 0\right) \\ P[\gamma_1 \frac{\partial \hat{f}(\underline{x}|\underline{\theta}_f)}{\partial \underline{\theta}_f} \underline{e}] & \text{if } \left(|\underline{\theta}_f|^2 = M_f \text{ and } \underline{e}^T \left(\frac{\partial \hat{f}}{\partial \underline{\theta}_f}\right)^T \underline{\theta}_f > 0\right) \end{cases} \quad (11.45)$$

$$\dot{\underline{\theta}}_g = \begin{cases} \gamma_2 \frac{\partial \hat{g}(\underline{x}|\underline{\theta}_g)}{\partial \underline{\theta}_g} \underline{e}u & \text{if } \left(|\underline{\theta}_g|^2 < M_g\right) \text{ or } \left(|\underline{\theta}_g|^2 = M_g \text{ and } \underline{e}^T \left(\frac{\partial \hat{g}}{\partial \underline{\theta}_g}\right)^T \underline{\theta}_g u \leq 0\right) \\ P[\gamma_2 \frac{\partial \hat{g}(\underline{x}|\underline{\theta}_g)}{\partial \underline{\theta}_g} \underline{e}u] & \text{if } \left(|\underline{\theta}_g|^2 = M_g \text{ and } \underline{e}^T \left(\frac{\partial \hat{g}}{\partial \underline{\theta}_g}\right)^T \underline{\theta}_g u > 0\right) \end{cases} \quad (11.46)$$

where γ_1 and γ_2 are positive constants, $P[*]$ is the projection operator

$$P\left[\gamma_1 \frac{\partial \hat{f}}{\partial \underline{\theta}_f} \underline{e}\right] \equiv \gamma_1 \frac{\partial \hat{f}}{\partial \underline{\theta}_f} \underline{e} - \gamma_1 \frac{\underline{e}^T \left(\frac{\partial \hat{f}}{\partial \underline{\theta}_f}\right)^T \underline{\theta}_f}{|\underline{\theta}_f|^2} \underline{\theta}_f \quad (11.47)$$

$$P\left[\gamma_2 \frac{\partial \hat{g}}{\partial \underline{\theta}_g} \underline{e}u\right] \equiv \gamma_2 \frac{\partial \hat{g}}{\partial \underline{\theta}_g} \underline{e}u - \gamma_2 \frac{\underline{e}^T \left(\frac{\partial \hat{g}}{\partial \underline{\theta}_g}\right)^T \underline{\theta}_g u}{|\underline{\theta}_g|^2} \underline{\theta}_g \quad (11.48)$$

174 Design & Stability Analysis of Fuzzy Identifiers of Nonlinear Dynamic Systems Ch. 11

$\underline{e} = \underline{x} - \hat{\underline{x}}$, and the initial parameters $\underline{\theta}_f(0)$ and $\underline{\theta}_g(0)$ are determined in the following way: for rules R_f^l and R_g^l in (11.35) and (11.36) that agree with the linguistic rules (11.17) and (11.18), set their initial parameters equal to the corresponding parameters in the linguistic rules. Otherwise, set the initial parameters arbitrarily. Therefore, as with the first-type fuzzy identifier, we incorporate linguistic descriptions about f and g into the fuzzy identifier by constructing the initial fuzzy identifier based on these linguistic descriptions.

The following theorem summarizes the properties of the second-type fuzzy identifier.

Theorem 11.3. The fuzzy identifier (11.16) with $\hat{f}$ and $\hat{g}$ given by (11.37) and (11.38) and the adaptive law (11.45) and (11.46) guarantees the following properties:

1. $|\underline{\theta}_f|^2 \le M_f$, $|\underline{\theta}_g|^2 \le M_g$, and $\hat{\underline{x}} \in L_\infty$.
2. there exist constants a and b such that

$$\int_0^t |\underline{e}(\tau)|^2 d\tau \le a + b \int_0^t |\underline{v}(\tau)|^2 d\tau \qquad (11.49)$$

where

$$\underline{v} \equiv \left[\frac{\partial \hat{f}(x|\underline{\theta}_{f0})}{\partial \underline{\theta}_f} - \frac{\partial \hat{f}(x|\underline{\theta}_f)}{\partial \underline{\theta}_f} \right]^T (\underline{\theta}_f^* - \underline{\theta}_f)$$

$$+ \left[\frac{\partial \hat{g}(x|\underline{\theta}_{g0})}{\partial \underline{\theta}_g} - \frac{\partial \hat{g}(x|\underline{\theta}_g)}{\partial \underline{\theta}_g} \right]^T (\underline{\theta}_g^* - \underline{\theta}_g)u + \underline{w} \qquad (11.50)$$

$\underline{\theta}_{f0} = \lambda_1 \underline{\theta}_f + (1-\lambda_1)\underline{\theta}_f^*$, $\underline{\theta}_{g0} = \lambda_2 \underline{\theta}_g + (1-\lambda_2)\underline{\theta}_g^*$ for some $\lambda_1, \lambda_2 \in [0,1]$, and $\underline{w}$, $\underline{\theta}_f^*$, $\underline{\theta}_g^*$ are defined in (11.13)–(11.15) with Θ_f^* and Θ_g^* replaced by $\underline{\theta}_f^*$ and $\underline{\theta}_g^*$.
3. if $\underline{v} \in L_2$, then $\lim_{t\to\infty}|\underline{e}(t)| = 0$.

Proof of this theorem is given in the Appendix.

Remark 11.4. Because we have much more freedom in choosing the parameters of this second-type fuzzy identifier than the first-type fuzzy identifier, the $\underline{w}$ for the second-type fuzzy identifier should be smaller than the first one. Therefore, the key factor that influences the value of $\underline{v}$ is the closeness of the $\underline{\theta}_f$ and $\underline{\theta}_g$ to their optimal values $\underline{\theta}_f^*$ and $\underline{\theta}_g^*$. From (11.50) we see that if $|\underline{\theta}_f^* - \underline{\theta}_f|$ and $|\underline{\theta}_g^* - \underline{\theta}_g|$ are small, then $\underline{v}$ will be small. Because the initial parameters $\underline{\theta}_f(0)$ and $\underline{\theta}_g(0)$ are determined based on the linguistic descriptions about f and g, good descriptions should make $\underline{\theta}_f(0)$ and $\underline{\theta}_g(0)$ close to $\underline{\theta}_f^*$ and $\underline{\theta}_g^*$ and therefore make $\underline{v}$ small.

Remark 11.5. By using the adaptive law (11.45) and (11.46), it is possible that some σ parameters are zero at certain points. To deal with this problem, we added a small $\epsilon > 0$ to the fuzzy logic systems (11.37) and (11.38) so that the fuzzy identifier is well defined for zero σ parameters. Another way to deal with this problem is to modify the adaptive law so that it is guaranteed that all the σ parameters are greater than a small $\epsilon > 0$. This can be achieved by using the projection operation. In fact, we used this strategy in Chapter 8. The approach here is yet another strategy.

11.5 APPLICATION TO THE CHAOTIC GLYCOLYTIC OSCILLATOR IDENTIFICATION

In this section, we simulate the two fuzzy identifiers for the following chaotic glycolytic oscillator [Holden, 19]

$$\dot{x}_1(t) = -x_1(t)x_2^2(t) + 0.999 + 0.42cos(1.75t) \quad (11.51)$$

$$\dot{x}_2(t) = x_1(t)x_2^2(t) - x_2(t) \quad (11.52)$$

Figure 11.2 shows the trajectory of this chaotic system in the phase plane from $t = 0$ to $t = 50$ with initial condition $x_1(0) = x_2(0) = 1.5$. For this system, we have $f_1(\underline{x}) = -x_1 x_2^2 + 0.999$, $f_2(\underline{x}) = x_1 x_2^2 - x_2$, $g_1(\underline{x}) = 0.42$, and $g_2(\underline{x}) = 0$. For simplicity, we assume that g_1 and g_2 are known; that is, we only consider the identification of f_1 and f_2. For each fuzzy identifier, we simulate two cases: (1) without linguistic descriptions about f_1 and f_2, and (2) with the following linguistic descriptions

$$R_f^1 : \text{IF } x_1 \text{ is near } 1 \text{ and } x_2 \text{ is near } 1,$$
$$\text{THEN } f_1 \text{ is near } 0, f_2 \text{ is near } 0 \quad (11.53)$$

$$R_f^2 : \text{IF } x_1 \text{ is near } 1 \text{ and } x_2 \text{ is near } 2,$$
$$\text{THEN } f_1 \text{ is near } -3, f_2 \text{ is near } 2 \quad (11.54)$$

$$R_f^3 : \text{IF } x_1 \text{ is near } 2 \text{ and } x_2 \text{ is near } 1,$$
$$\text{THEN } f_1 \text{ is near } -1, f_2 \text{ is near } 1 \quad (11.55)$$

where *near c*, $c = 1, 0, 2, -3,$ or -1, is a fuzzy set *nc* with membership function $\mu_{nc}(z) = exp[-\frac{1}{2}(\frac{z-c}{0.25})^2]$. $R_f^1 - R_f^3$ are obtained by evaluating f_1 and f_2 at three points $\underline{x} = (1, 1)^T, (1, 2)^T,$ and $(2, 1)^T$, and then fuzzifying the $(\underline{x}, f(\underline{x}))$ pairs using the nonsingleton fuzzifier in Section 2.4. For practical systems, this kind of information is provided by human experts who are familiar with the behavior of the systems. In all the simulations in the sequel, we chose $\alpha = 1$, $\gamma_1 = 4$, $M_f = 10^6$, and $\epsilon = 10^{-6}$, and solved the differential equations using the MATLAB command "ode23" which uses the second-order and third-order Runge-Kutta method.

To simulate the first-type fuzzy identifier, we first need to define some fuzzy sets to cover the state space. From Figure 11.2 we see that most values of x_1

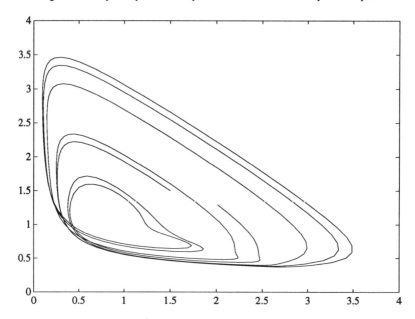

Figure 11.2 Trajectory of the chaotic glycolytic oscillator in the phase plane from $t_0 = 0$ to $t_f = 50$ with initial condition $x_1(0) = x_2(0) = 1.5$.

and x_2 are in the interval $[0.5, 3.5]$. Therefore, we define the following seven fuzzy sets F^j ($j = 1, 2, \ldots, 7$) for both x_1 and x_2 (that is, $m_1 = m_2 = 7$): $\mu_{F^j}(x_i) = exp[-\frac{1}{2}(\frac{x_i - \bar{x}^j}{0.25})^2]$, where $\bar{x}^j = 0.5 \times j$, $j = 1, 2, \ldots, 7$, and $i = 1, 2$. Clearly, these fuzzy sets include the fuzzy sets in the IF parts of $R_f^1 - R_f^3$. For this choice of F^j, we have 49 fuzzy rules to construct each $\hat{f}_i$ ($i = 1, 2$); that is, each $\hat{f}_i$ has 49 free parameters. We simulated two cases: (1) without incorporating the $R_f^1 - R_f^3$, that is, all elements of $\Theta_f(0)$ were chosen randomly in the interval $[-2, 2]$; and (2) incorporating the $R_f^1 - R_f^3$, that is, the elements of $\Theta_f(0)$ that correspond to the conditions in the IF parts of $R_f^1 - R_f^3$ were chosen according to the corresponding THEN parts of $R_f^1 - R_f^3$, and other elements were still chosen randomly in the interval $[-2, 2]$. Figures 11.3 and 11.4 show the states $x_1(t)$ and $x_2(t)$ with the corresponding $\hat{x}_1(t)$ and $\hat{x}_2(t)$, respectively, for the first case, and Figures 11.5 and 11.6 show the same results for the second case. Comparing Figures 11.3 and 11.5 and Figures 11.4 and 11.6, we see that the adaptation speed and accuracy were greatly improved by incorporating the linguistic rules $R_f^1 - R_f^3$.

We simulated the second-type fuzzy identifier for the same two cases: (1) without incorporating the $R_f^1 - R_f^3$, that is, all elements of $\underline{\theta}_f(0)$ were chosen randomly in the interval $[-2, 2]$; and (2) incorporating the $R_f^1 - R_f^3$, that is, some elements of $\underline{\theta}_f(0)$ were chosen according to the fuzzy sets in $R_f^1 - R_f^3$, and the remaining elements were chosen randomly in the interval $[-2, 2]$. We chose $N = 10$, so that on average each $\hat{f}_i$ ($i = 1, 2$) contains 40 free parameters (see (11.37)).

Sec 11.5 Application to the Chaotic Glycolytic Oscillator Identification

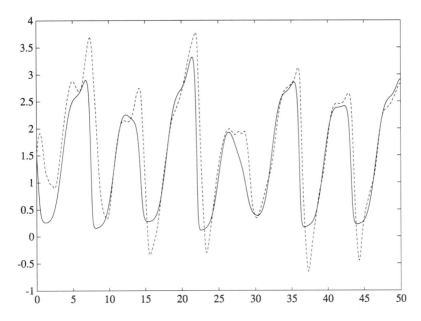

Figure 11.3 $x_1(t)$ (solid line) and $\hat{x}_1(t)$ (dashed line) using the first-type fuzzy identifier without incorporating any linguistic information.

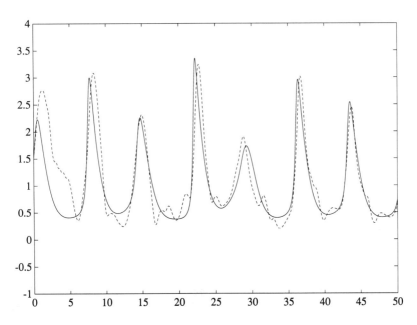

Figure 11.4 $x_2(t)$ (solid line) and $\hat{x}_2(t)$ (dashed line) using the first-type fuzzy identifier without incorporating any linguistic information.

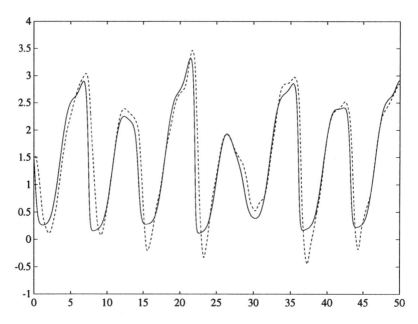

Figure 11.5 $x_1(t)$ (solid line) and $\hat{x}_1(t)$ (dashed line) using the first-type fuzzy identifier after incorporating the linguistic descriptions (11.53)–(11.55).

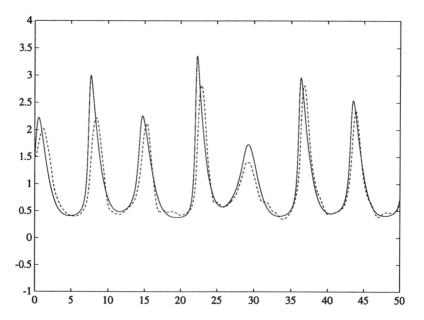

Figure 11.6 $x_2(t)$ (solid line) and $\hat{x}_2(t)$ (dashed line) using the first-type fuzzy identifier after incorporating the linguistic descriptions (11.53)–(11.55).

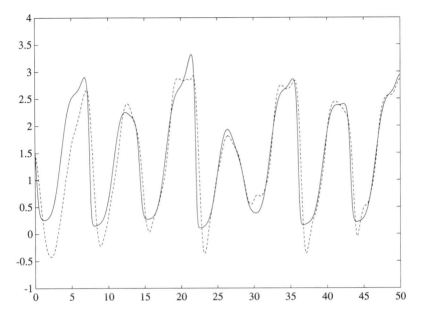

Figure 11.7 $x_1(t)$ (solid line) and $\hat{x}_1(t)$ (dashed line) using the second-type fuzzy identifier without incorporating any linguistic information.

Figures 11.7 and 11.8 show the $x_1(t), \hat{x}_1(t)$ and $x_2(t), \hat{x}_2(t)$ curves for the first case, respectively, and Figures 11.9 and 11.10 show the same results for the second case. Comparing Figures 11.7 and 11.9 and Figures 11.8 and 11.10, we see that the adaptation speed and accuracy were greatly improved by incorporating the linguistic rules $R_f^1 - R_f^3$.

Comparing Figures 11.3–11.6 with the corresponding Figures 11.7–11.10, we see that although the second-type fuzzy identifier used less free parameters than the first-type fuzzy identifier, its performance is better. This may be caused by the following two reasons: (1) although each fuzzy logic system in the first-type fuzzy identifier has 49 free parameters, some of them were never used because from Figure 11.2 we see that the states of the system never enter some regions in the state space so that the parameters responsible for these regions have no influence on the performance of the identifier. That is, the number of actually useful free parameters of the second-type fuzzy identifier is more than that of the first-type fuzzy identifier; and, (2) the first-type fuzzy identifier is linear in the parameter so that in principle it should converge to a single solution, whereas the second-type fuzzy identifier is nonlinear in the parameter so that it has a larger functional space to search. In principle, the second-type fuzzy identifier has the danger of being trapped at a local minimum, but this seems not to happen in our simulations, especially when we incorporated the linguistic rules.

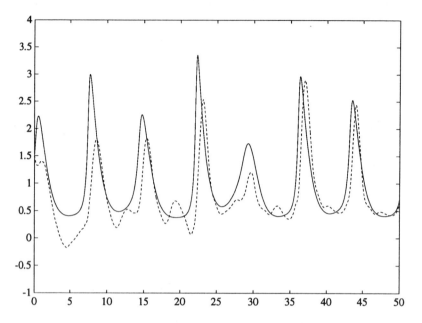

Figure 11.8 $x_2(t)$ (solid line) and $\hat{x}_2(t)$ (dashed line) using the second-type fuzzy identifier without incorporating any linguistic information.

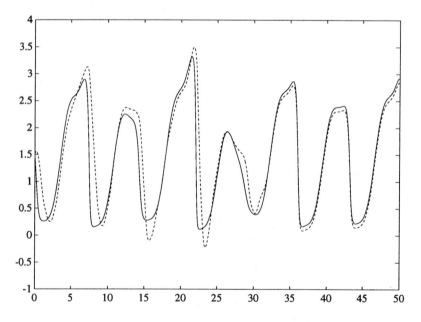

Figure 11.9 $x_1(t)$ (solid line) and $\hat{x}_1(t)$ (dashed line) using the second-type fuzzy identifier after incorporating the linguistic descriptions (11.53)–(11.55).

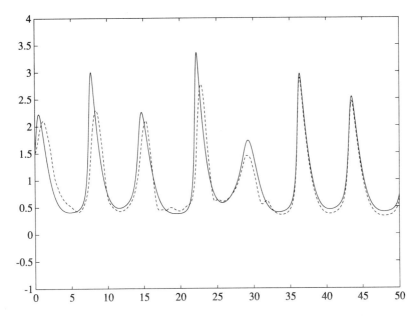

Figure 11.10 $x_2(t)$ (solid line) and $\hat{x}_2(t)$ (dashed line) using the second-type fuzzy identifier after incorporating the linguistic descriptions (11.53)–(11.55).

11.6 CONCLUDING REMARKS

In this chapter two identifiers of nonlinear dynamic systems were developed based on the fuzzy system models. We proved that all signals in the fuzzy identifiers are uniformly bounded and provided conditions under which the identification errors converge to zero. We also proved that the fuzzy identifiers are capable of following the output of a very general nonlinear dynamic system to arbitrary accuracy in any finite time interval. The most important advantage of the fuzzy identifiers is that linguistic descriptions about the systems (in terms of fuzzy IF-THEN rules) can be directly incorporated into the fuzzy identifiers. We simulated the two fuzzy identifiers for the chaotic glycolytic oscillator. The results show that (1) they could identify the chaotic system at a reasonable speed and accuracy without using any linguistic information, and (2) by incorporating some linguistic information, the speed and accuracy of the fuzzy identifiers were greatly improved.

12
FUZZY ADAPTIVE FILTERS

12.1 INTRODUCTION

Filters are information processors. In practice, information usually comes from two sources: sensors which provide numerical data associated with a problem, and human experts who provide linguistic descriptions (often in the form of fuzzy IF-THEN rules) about the problem. Existing filters can only process numerical data, whereas existing expert systems can only make use of linguistic information. Therefore, their successful applications are limited to problems where either linguistic rules or numerical data do not play a critical role. There are, however, a large number of practical problems in economics, seismology, management, and so on, where both linguistic and numerical information are critical. At present, when we are faced with such problems, we use linguistic information, consciously or unconsciously, in the choice among different filters, the evaluation of filter performance, the choice of filter orders, the interpretation of filtering results, and so on. There are serious limitations to using linguistic information in this way because for most practical problems the linguistic information (in its natural form) is not about which kind of filter should be chosen or what the order of the filter should be, but is in the form of fuzzy IF-THEN rules. The purpose of this chapter is to develop new kinds of nonlinear adaptive filters, which we refer to as *fuzzy adaptive filters*, that make use of both linguistic and numerical information in their natural form.

A fuzzy adaptive filter is constructed from a set of changeable fuzzy IF-THEN rules. These fuzzy rules come either from human experts or by matching input-output pairs through an adaptation procedure. The adaptive algorithms update the parameters of the membership functions which characterize the fuzzy concepts

Sec. 12.2 RLS Fuzzy Adaptive Filter

in the IF-THEN rules by minimizing some criterion functions. Two fuzzy adaptive filters are developed in this chapter which use recursive least squares (RLS) and least mean squares (LMS) algorithms, respectively.

12.2 RLS FUZZY ADAPTIVE FILTER

Our RLS fuzzy adaptive filter solves the following problem.

Problem 12.1. Consider a real-valued vector sequence $[\underline{x}(k)]$ and a real-valued scalar sequence $[d(k)]$, where $k = 0, 1, 2, \ldots$ is the time index, and $\underline{x}(k) \in U \equiv [C_1^-, C_1^+] \times [C_2^-, C_2^+] \times \cdots \times [C_n^-, C_n^+] \subset R^n$ (we call U and R the input and output spaces of the filter, respectively). At each time point k, we are given the values of $\underline{x}(k)$ and $d(k)$. The problem is: at each time point $k = 0, 1, 2, \ldots$, determine an adaptive filter $f_k : U \subset R^n \to R$ such that

$$J(k) = \sum_{i=0}^{k} \lambda^{k-i} [d(i) - f_k(\underline{x}(i))]^2 \tag{12.1}$$

is minimized, where $\lambda \in (0, 1]$ is a forgetting factor.

The preceding problem is quite general. As a particular example, consider the following time-series prediction problem: we measure the values of a bounded time series $[y(k)]$ at each time point $k = 0, 1, 2, \ldots$; at time point $k - 1$, we want to determine a filter $f_{k-1} : U \to R$ such that $f_{k-1}[y(k-1), \ldots, y(k-n)]$ is an optimal prediction of $y(k)$ in some sense. For this problem, we have $\underline{x}(k) = [y(k-1), \ldots, y(k-n)]^T$, $d(k) = y(k)$, and the "in some sense" means to minimize the $J(k)$ of (12.1). If we constrain the f_k's to be linear functions, the problem becomes an FIR adaptive filter design problem [Cowan and Grant, 11; Widrow and Stearns, 96]. If the f_k's are Volterra series expansions, we have an adaptive polynomial filter design problem [Mathews, 43]. If the f_k's are multilayer perceptrons or radial basis function expansions, the problem becomes the neural nets adaptive filter design problem [Chen, Gibson, Cowan, and Grant, 7; Chen, Gibson, Cowan, and Grant, 8].

Design Procedure of the RLS Fuzzy Adaptive Filter

Step 1:

Define m_i fuzzy sets in each interval $[C_i^-, C_i^+]$ of the input space U, which are labeled as F_i^{ji} ($i = 1, 2, \ldots, n$, $ji = 1, 2, \ldots, m_i$, note that ji is a single index), in the following way: the m_i membership functions $\mu_{F_i^{ji}}$ cover the interval $[C_i^-, C_i^+]$, in the sense that for each $x_i \in [C_i^-, C_i^+]$, there exists at least one $\mu_{F_i^{ji}}(x_i) \neq 0$.

Step 2:

Construct a set of $\prod_{i=1}^{n} m_i$ fuzzy IF-THEN rules in the following form:

$$R^{(j1,\ldots,jn)} : IF\ x_1\ is\ F_1^{j1}\ and\ \cdots\ and\ x_n\ is\ F_n^{jn},\ THEN\ d\ is\ G^{(j1,\ldots,jn)} \tag{12.2}$$

where $\underline{x} = (x_1, \ldots, x_n)^T \in U$ (the filter input), $d \in R$ (the filter output), $ji = 1, 2, \ldots, m_i$ with $i = 1, 2, \ldots, n$, F_i^{ji}'s are the same labels of the fuzzy sets defined in Step 1, and the $G^{(j1,\ldots,jn)}$'s are labels of fuzzy sets defined in the output space which are determined in the following way: if there are linguistic rules from human experts in the form of (12.2), set $G^{(j1,\ldots,jn)}$ to be the corresponding linguistic terms of these rules. Otherwise, set $\mu_{G^{(j1,\ldots,jn)}}$ to be an arbitrary membership function over the output space R. It is in this way that we incorporate linguistic rules into the fuzzy adaptive filter, that is, we use linguistic rules to construct the initial filter.

Step 3:

Construct the filter f_k based on the $\prod_{i=1}^{n} m_i$ rules in Step 2 as follows:

$$f_k(\underline{x}) = \frac{\sum_{j1=1}^{m_1} \cdots \sum_{jn=1}^{m_n} \theta^{(j1,\ldots,jn)} (\mu_{F_1^{j1}}(x_1) \cdots \mu_{F_n^{jn}}(x_n))}{\sum_{j1=1}^{m_1} \cdots \sum_{jn=1}^{m_n} (\mu_{F_1^{j1}}(x_1) \cdots \mu_{F_n^{jn}}(x_n))} \quad (12.3)$$

where $\underline{x} = (x_1, \ldots, x_n)^T \in U$, $\mu_{F_i^{ji}}$'s are membership functions defined in Step 1, and $\theta^{(j1,\ldots,jn)} \in R$ is the point at which $\mu_{G^{(j1,\ldots,jn)}}$ achieves its maximum value. Due to the way in which we defined the $\mu_{F_i^{ji}}$'s in Step 1, the denominator of (12.3) is nonzero for all the points of U; therefore, the filter f_k of (12.3) is well defined. Equation (12.3) is obtained by combining the $\prod_{i=1}^{n} m_i$ rules of Step 2 using product inference and center average defuzzification (see Lemma 2.1). Another way of interpreting (12.3) is as follows. For a given input $\underline{x} \in U$, we determine the filter output $f_k(\underline{x})$ as a weighted average of the $\prod_{i=1}^{n} m_i$ points $\theta^{(j1,\ldots,jn)}$ in the output space at which the fuzzy sets $G^{(j1,\ldots,jn)}$ of the THEN parts of the $\prod_{i=1}^{n} m_i$ rules have maximum membership values, and the weight $\mu_{F_1^{j1}}(x_1) \cdots \mu_{F_n^{jn}}(x_n)$ for $\theta^{(j1,\ldots,jn)}$ is proportional to the membership values for which $\underline{x}$ satisfies the IF part of $R^{(j1,\ldots,jn)}$. This is a reasonable filter because $\theta^{(j1,\ldots,jn)}$'s are the "most likely" points in the output space based on the $\prod_{i=1}^{n} m_i$ rules, and the point $\theta^{(j1,\ldots,jn)}$ should be given more weight if the given input point $\underline{x}$ satisfies the corresponding IF part "more likely" (in the sense of larger membership value).

In (12.3), the weights $\mu_{F_1^{j1}}(x_1) \cdots \mu_{F_n^{jn}}(x_n)$ are fixed functions of $\underline{x}$; therefore, the free design parameters of the fuzzy adaptive filter are the $\theta^{(j1,\ldots,jn)}$'s which are now collected as a $\prod_{i=1}^{n} m_i$-dimensional vector

$$\underline{\theta} \equiv (\theta^{(1,1,\ldots,1)}, \ldots, \theta^{(m_1,1,\ldots,1)}; \theta^{(1,2,1,\ldots,1)}, \ldots, \theta^{(m_1,2,1,\ldots,1)}; \ldots;$$

$$\theta^{(1,m_2,1,\ldots,1)}, \ldots, \theta^{(m_1,m_2,1,\ldots,1)}; \ldots; \theta^{(1,m_2,\ldots,m_n)}, \ldots, \theta^{(m_1,m_2,\ldots,m_n)})^T \quad (12.4)$$

Define the *fuzzy basis functions* (Chapter 4)

$$p^{(j1,\ldots,jn)}(\underline{x}) = \frac{\mu_{F_1^{j1}}(x_1) \cdots \mu_{F_n^{jn}}(x_n)}{\sum_{j1=1}^{m_1} \cdots \sum_{jn=1}^{m_n} (\mu_{F_1^{j1}}(x_1) \cdots \mu_{F_n^{jn}}(x_n))} \quad (12.5)$$

and collect them as a $\prod_{i=1}^{n} m_i$-dimensional vector $\underline{p}(\underline{x})$ in the same ordering as the $\underline{\theta}$ of (12.4).

$$\underline{p}(\underline{x}) \equiv (p^{(1,1,\ldots,1)}(\underline{x}), \ldots, p^{(m_1,1,\ldots,1)}(\underline{x}); p^{(1,2,1,\ldots,1)}(\underline{x}), \ldots, p^{(m_1,2,1,\ldots,1)}(\underline{x});$$

$$\ldots; p^{(1,m_2,1,\ldots,1)}(\underline{x}), \ldots, p^{(m_1,m_2,1,\ldots,1)}(\underline{x}); \ldots; p^{(1,m_2,\ldots,m_n)}(\underline{x}),$$

$$\ldots, p^{(m_1,m_2,\ldots,m_n)}(\underline{x}))^T \quad (12.6)$$

Sec. 12.2 RLS Fuzzy Adaptive Filter

Based on (12.4) and (12.6) we can now rewrite (12.3) as

$$f_k(\underline{x}) = \underline{p}^T(\underline{x})\underline{\theta} \qquad (12.7)$$

We see from (12.7) that f_k is linear in the parameter; therefore, we can use the fast-convergent RLS algorithm to update the parameters $\underline{\theta}$.

Step 4:

Use the following RLS algorithm [Cowan and Grant, 11] to update $\underline{\theta}$. Let the initial estimate of $\underline{\theta}$, $\underline{\theta}(0)$ be determined as in Step 2, and $P(0) = \sigma I$, where σ is a small positive constant, and I is the $\prod_{i=1}^{n} m_i - by - \prod_{i=1}^{n} m_i$ identity matrix. At each time point $k = 1, 2, \ldots$, do the following:

$$\underline{\phi}(k) = \underline{p}(\underline{x}(k)) \qquad (12.8)$$

$$P(k) = \frac{1}{\lambda}[P(k-1) - P(k-1)\underline{\phi}(k)(\lambda + \underline{\phi}^T(k)P(k-1)\underline{\phi}(k))^{-1}\underline{\phi}^T(k)P(k-1)] \qquad (12.9)$$

$$K(k) = P(k-1)\underline{\phi}(k)[\lambda + \underline{\phi}^T(k)P(k-1)\underline{\phi}(k)]^{-1} \qquad (12.10)$$

$$\underline{\theta}(k) = \underline{\theta}(k-1) + K(k)(d(k) - \underline{\phi}^T(k)\underline{\theta}(k-1)) \qquad (12.11)$$

where $[\underline{x}(k)]$ and $[d(k)]$ are the sequences defined previously in Problem 12.1, $\underline{p}(*)$ is defined in (12.6), and λ is the forgetting factor in (12.1).

Some comments on this RLS fuzzy adaptive filter are now in order.

Remark 12.1. The RLS algorithm (12.9)–(12.11) is obtained by minimizing $J(k)$ of (12.1) with f_k constrained to be in the form of (12.7). Because f_k of (12.7) is linear in the parameter, the derivation of (12.9)–(12.11) is the same as that of the FIR linear adaptive filter [Cowan and Grant, 11]. Therefore, we omit the details.

Remark 12.2. The RLS algorithm (12.9)–(12.11) can be viewed as updating the $\prod_{i=1}^{n} m_i$ rules in the form of (12.2) by changing the "centers" $\theta^{(j1,\ldots,jn)}$ of the THEN parts of these rules in the direction of minimizing the criterion function (12.1). We are allowed only to change these "centers." The membership functions $\mu_{F_i^{ji}}$ of the IF parts of the rules are fixed at the very beginning and are not allowed to change. Therefore, a good choice of $\mu_{F_i^{ji}}$'s is important to the success of the entire filter. In the next section, we will allow the $\mu_{F_i^{ji}}$'s also to change during the adaptation procedure.

Remark 12.3. It was proven in Chapter 2 that functions in the form of (12.3) are universal approximators. That is, for any real continuous function g on the compact set U, there exists a function in the form of (12.3) such that it can uniformly approximate g over U to arbitrary accuracy. Consequently, our fuzzy adaptive filter is a powerful nonlinear adaptive filter in the sense that it has the capability of performing difficult nonlinear filtering operations.

Remark 12.4. The fuzzy adaptive filter (12.7) performs a two-stage operation on the input vector $\underline{x}$. First, it performs a nonlinear transformation $\underline{p}(*)$ on $\underline{x}$; then the filter output is obtained as a linear combination of these transformed signals. In this sense, our fuzzy adaptive filter is similar to the radial basis function [Chen, Gibson, Cowan, and Grant, 8; Powell, 58] and potential function [Meisel, 44] approaches. The unique feature of our fuzzy adaptive filter, which is not shared by other nonlinear adaptive filters, is that linguistic rules can be incorporated into the filter, as discussed next.

Remark 12.5. Linguistic information (in the form of the fuzzy IF-THEN rules of (12.2)) and numerical information (in the form of desired input-output pairs $(\underline{x}(k), d(k))$) are combined into the filter in the following way. Due to Steps 2–4, linguistic IF-THEN rules are directly incorporated into the filter (12.3) by constructing the initial filter based on the linguistic rules. Also, due to the adaptation of Step 4, numerical pairs $(\underline{x}(k), d(k))$ are incorporated into the filter by updating the filter parameters such that the filter output "matches" the pairs in the sense of minimizing (12.1). It is natural and reasonable to assume that linguistic information from human experts is provided in the form of (12.2) because the rules of (12.2) state what the filter outputs should be in some input situations, where "what should be" and "some situations" are represented by linguistic terms which are characterized by fuzzy membership functions. On the other hand, it is obvious that the most natural form of numerical information is provided in the form of input-output pairs $(\underline{x}(k), d(k))$.

Remark 12.6. By fixing the fuzzy membership functions on the input space U at the very beginning, we obtained a nonlinear filter which is linear in the parameter. Therefore, we could use the fast-convergent RLS algorithm in the adaptation procedure. The price paid is that we had to include all the $\prod_{i=1}^{n} m_i$ possible rules in the filter because if a region of U is not covered by any rules and an input $\underline{x}$ to the filter happens to be in this region, then the filter response will be very poor. As a result, for problems of high-dimension n and large m_i, the computations involved in this fuzzy adaptive filter are intense because at each time point k we need to perform the $\prod_{i=1}^{n} m_i$-dimensional matrix-to-vector multiplications of (12.9)–(12.11) and to evaluate the values of the $\prod_{i=1}^{n} m_i$ fuzzy basis functions of (12.8) (see also (12.6) and (12.5)). Although these computations are highly parallelizable, we may not be able to use the filter in some practical situations where computing power is limited. Therefore, we will develop another fuzzy adaptive filter which involves much less computation next.

12.3 LMS FUZZY ADAPTIVE FILTER

Our LMS fuzzy adaptive filter solves the following problem.

Problem 12.2. Consider the same input sequence $[\underline{x}(k)]$ and output sequence $[d(k)]$ as in Problem 12.1. The problem is: at each time point $k = 1, 2, \ldots,$

Sec. 12.3 LMS Fuzzy Adaptive Filter

determine an adaptive filter $f_k : U \to R$ such that

$$L = E[(d(k) - f_k(\underline{x}(k)))^2] \tag{12.12}$$

is minimized.

Design Procedure of the LMS Fuzzy Adaptive Filter

Step 1:

Define M fuzzy sets F_i^l in each interval $[C_i^-, C_i^+]$ of U with the following Gaussian membership functions

$$\mu_{F_i^l}(x_i) = exp\left[-\frac{1}{2}\left(\frac{x_i - \bar{x}_i^l}{\sigma_i^l}\right)^2\right] \tag{12.13}$$

where $l = 1, 2, \ldots, M$, $i = 1, 2, \ldots, n$, $x_i \in [C_i^-, C_i^+]$, and $\bar{x}_i^l$ and σ_i^l are free parameters which will be updated in the LMS adaptation procedure of Step 4.

Step 2:

Construct a set of M (in general, $M << \prod_{i=1}^{n} m_i$) fuzzy IF-THEN rules in the following form:

$$R^l : \text{ IF } x_1 \text{ is } F_1^l \text{ and } \cdots \text{ and } x_n \text{ is } F_n^l, \text{ THEN } d \text{ is } G^l \tag{12.14}$$

where $\underline{x} = (x_1, \ldots, x_n)^T \in U$, $d \in R$, F_i^l's are defined in Step 1, and G^l's are fuzzy sets defined in R which are determined as follows: *if there are linguistic rules in the form of (12.14), set F_i^l's and G^l to be the labels of these linguistic rules; otherwise, choose μ_{G^l} and the parameters $\bar{x}_i^l$ and σ_i^l arbitrarily*. The (parameters of) membership functions $\mu_{F_i^l}$ and μ_{G^l} in these rules will change during the LMS adaptation procedure of Step 4. Therefore, the rules constructed in this step are initial rules of the fuzzy adaptive filter. As in the RLS fuzzy adaptive filter, we incorporate linguistic rules into the LMS fuzzy adaptive filter by constructing the initial filter based on these rules.

Step 3:

Construct the filter $f_k : U \to R$ based on the M rules of Step 2 as follows:

$$f_k(\underline{x}) = \frac{\sum_{l=1}^{M} \theta^l \left(\prod_{i=1}^{n} \mu_{F_i^l}(x_i)\right)}{\sum_{l=1}^{M} \left(\prod_{i=1}^{n} \mu_{F_i^l}(x_i)\right)} \tag{12.15}$$

where $\underline{x} = (x_1, \ldots, x_n)^T \in U$, $\mu_{F_i^l}$'s are the Gaussian membership functions of (12.13), and $\theta^l \in R$ is any point at which μ_{G^l} achieves its maximum value. The filter (12.15) is constructed in the same way as (12.3) and shares the same interpretation. Because we chose the membership functions $\mu_{F_i^l}(x_i)$ to be Gaussian functions which are nonzero for any $x_i \in [C_i^-, C_i^+]$, the denominator of (12.15) is nonzero for any $\underline{x} \in U$. Therefore, the filter f_k of (12.15) is well defined. Because the θ^l as well as $\bar{x}_i^l$ and σ_i^l are free parameters, the filter (12.15) is nonlinear in the parameters.

Step 4:

Use the following LMS algorithm [Widrow and Stearns, 96] to update the filter parameters θ^l, $\bar{x}_i^l$, and σ_i^l: let the initial $\theta^l(0)$, $\bar{x}_i^l(0)$, and $\sigma_i^l(0)$ be as determined in Step 2; at each time point $k = 1, 2, \ldots$, do the following:

$$\theta^l(k) = \theta^l(k-1) + \alpha[d(k) - f_k]\frac{a^l(k-1)}{b(k-1)} \qquad (12.16)$$

$$\bar{x}_i^l(k) = \bar{x}_i^l(k-1) + \alpha[d(k) - f_k]\frac{\theta^l(k-1) - f_k}{b(k-1)} a^l(k-1) \frac{x_i(k) - \bar{x}_i^l(k-1)}{(\sigma_i^l(k-1))^2} \qquad (12.17)$$

$$\sigma_i^l(k) = \sigma_i^l(k-1) + \alpha[d(k) - f_k]\frac{\theta^l(k-1) - f_k}{b(k-1)} a^l(k-1) \frac{(x_i(k) - \bar{x}_i^l(k-1))^2}{(\sigma_i^l(k-1))^3} \qquad (12.18)$$

where $a^l(k-1) = \prod_{i=1}^n \exp[-\frac{1}{2}(\frac{x_i(k) - \bar{x}_i^l(k-1)}{\sigma_i^l(k-1)})^2]$, $b(k-1) = \sum_{l=1}^M a^l(k-1)$, $f_k = \frac{\sum_{l=1}^M \theta^l a^l(k-1)}{b(k-1)}$, α is a small positive stepsize, $l = 1, 2, \ldots, M$, and $i = 1, 2, \ldots, n$. Equations (12.16)–(12.18) are obtained by taking the gradient of L (12.12) (ignore the expectation E) with respect to the parameters and using the specific formula of (12.15) and (12.13).

Some comments on this LMS fuzzy adaptive filter are now in order.

Remark 12.7. From Steps 2–4 we see that the initial LMS fuzzy adaptive filter is constructed based on linguistic rules from human experts and some arbitrary rules (in the sense that the parameters of membership functions $\mu_{F_i^l}$ and μ_{G^l} which characterize these rules are chosen arbitrarily). Both sets of rules are updated during the LMS adaptation procedure of Step 4 by changing the parameters in the direction of minimizing the L of (12.12). Because minimizing (12.12) can be viewed as matching the input-output pairs $[\underline{x}(k); d(k)]$, our LMS fuzzy adaptive filter combines both linguistic and numerical information in its design.

Remark 12.8. Because the LMS algorithm is a gradient algorithm, a good choice of initial parameters is very important to its convergence. Because we use linguistic information to choose the initial parameters, the adaptation procedure should converge quickly if the linguistic rules provide good instructions for how the filter should perform, that is, good descriptions of the input-ouput pairs $[\underline{x}(k); d(k)]$. Therefore, although LMS algorithms in general are slow to converge, our LMS algorithm in particular may converge fast, provided that there are sufficient linguistic rules.

Remark 12.9. The filter f_k of (12.15) can match any input-output pair $[\underline{x}(k); d(k)]$ to arbitrary accuracy by properly choosing the parameters θ^l, $\bar{x}_i^l$ and σ_i^l, as we show next. For given $[\underline{x}(k); d(k)]$, let $\bar{x}_i^1 = x_i(k)$, $\bar{x}_i^l \neq x_i(k)$ for $l \neq 1$, and $\theta^1 = d(k)$; therefore, for $\underline{x} = \underline{x}(k)$ in (12.15), the weight $(\prod_{i=1}^n \mu_{F_i^1}(x_i(k)))$ for $\theta^1 = d(k)$ equals 1 for any choice of σ_i^1, and the other $M - 1$ weights $(\prod_{i=1}^n \mu_{F_i^l}(x_i(k)))$ for θ^l with $l \neq 1$ can be arbitrarily close to zero if we choose all σ_i^l to be sufficiently small (see (12.13) and notice that $x_i(k) \neq \bar{x}_i^l$ for $l \neq 1$). As a result, $|d(k) -$

$f_k(\underline{x}(k))|$ can be arbitrarily small. Because of this property and the freedom to update the parameters during the adaptation procedure, we can hope that we have a well-performing filter using only a small number of rules (that is, we may choose $M << \prod_{i=1}^{n} m_i$).

12.4 APPLICATION TO NONLINEAR CHANNEL EQUALIZATION

Nonlinear distortion over a communication channel is now a significant factor hindering further increase in the attainable data rate in high-speed data transmission [Biglieri, Gersho, Gitlin, and Lim, 3; Falconer, 14]. Because the received signal over a nonlinear channel is a nonlinear function of the past values of the transmitted symbols, and the nonlinear distortion varies with time and from place to place, effective equalizers for nonlinear channels should be nonlinear and adaptive.

In Biglieri, Gersho, Gitlin, and Lim [3] and Falconer [14], polynomial adaptive filters were developed for nonlinear channel equalization. In Chen, Gibson, Cowan, and Grant [7] and Chen, Gibson, Cowan, and Grant [8], multilayer perceptrons and radial basis function expansions were used as adaptive equalizers for nonlinear channels. Because nonlinear channels include a very broad spectrum of nonlinear distortion, it is difficult to say which nonlinear adaptive filter is dominantly better than the others. Therefore, it is worth trying other new nonlinear structures as prototypes of nonlinear adaptive filters in addition to the existing Volterra series, multilayer perceptron, radial basis function expansions, and so on. The RLS and LMS adaptive filters are examples of such new nonlinear adaptive filters. In this section, we use them as equalizers for nonlinear channels.

The digital communication system considered here is shown in Figure 12.1, where the *channel* includes the effects of the transmitter filter, the transmission medium, the receiver matched filter, and other components. The transmitted data sequence $s(k)$ is assumed to be an independent sequence taking values from $\{-1, 1\}$ with equal probability. The inputs to the equalizer, $x(k), x(k-1), \cdots, x(k-n+1)$, are the channel outputs corrupted by an additive noise $e(k)$. The task of the equalizer at the sampling instant k is to produce an estimate of the transmitted symbol $s(k-d)$ using the information contained in $x(k), x(k-1), \cdots, x(k-n+1)$, where the integers n and d are known as the order and the lag of the equalizer, respectively.

We use the geometric formulation of the equalization problem due to Chen, Gibson, Cowan, and Grant [7] and Chen, Gibson, Cowan, and Grant [8]. Using similar notation to that in Chen, Gibson, Cowan, and Grant [7] and Chen, Gibson, Cowan, and Grant [8], define

$$P_{n,d}(1) = \{\hat{\underline{x}}(k) \in R^n | s(k-d) = 1\} \quad (12.19)$$

$$P_{n,d}(-1) = \{\hat{\underline{x}}(k) \in R^n | s(k-d) = -1\} \quad (12.20)$$

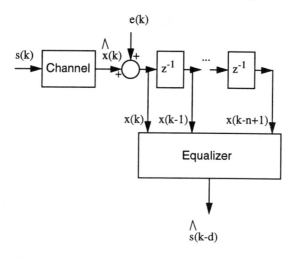

Figure 12.1 Schematic of data transmission system.

where

$$\hat{\underline{x}}(k) = [\hat{x}(k), \hat{x}(k-1), \cdots, \hat{x}(k-n+1)]^T \quad (12.21)$$

$\hat{x}(k)$ is the noise-free output of the channel (see Figure 12.1), and $P_{n,d}(1)$ and $P_{n,d}(-1)$ represent the two sets of possible channel noise-free output vectors $\hat{\underline{x}}(k)$ that can be produced from sequences of the channel inputs containing $s(k-d) = 1$ and $s(k-d) = -1$, respectively. The equalizer can be characterized by the function

$$g_k : R^n \to \{-1, 1\} \quad (12.22)$$

with

$$\hat{s}(k-d) = g_k(\underline{x}(k)) \quad (12.23)$$

where

$$\underline{x}(k) = [x(k), x(k-1), \cdots, x(k-n+1)]^T \quad (12.24)$$

is the observed channel output vector. Let $p_1[\underline{x}(k)|\hat{\underline{x}}(k) \in P_{n,d}(1)]$ and $p_{-1}[\underline{x}(k)|\hat{\underline{x}}(k) \in P_{n,d}(-1)]$ be the conditional probability density functions of $\underline{x}(k)$ given $\hat{\underline{x}}(k) \in P_{n,d}(1)$ and $\hat{\underline{x}}(k) \in P_{n,d}(-1)$, respectively. It was shown in the works of Chen, Gibson, Cowan, and Grant [7, 8] that the equalizer which is defined by

$$f_{opt}(\underline{x}(k)) = sgn[p_1(\underline{x}(k)|\hat{\underline{x}}(k) \in P_{n,d}(1)) - p_{-1}(\underline{x}(k)|\hat{\underline{x}}(k) \in P_{n,d}(-1))] \quad (12.25)$$

achieves the minimum bit error rate for the given order n and lag d, where $sgn(y) = 1(-1)$ if $y \geq 0$ ($y < 0$). If the noise $e(k)$ is zero-mean and Gaussian with covariance matrix

$$Q = E[(e(k), \ldots, e(k-n+1))(e(k), \ldots, e(k-n+1))^T] \quad (12.26)$$

Sec. 12.4 Application to Nonlinear Channel Equalization 191

then from $x(k) = \hat{x}(k) + e(k)$ we have that

$$p_1\left[\underline{x}(k)|\underline{\hat{x}}(k) \in P_{n,d}(1)\right] - p_{-1}\left[\underline{x}(k)|\underline{\hat{x}}(k) \in P_{n,d}(-1)\right]$$

$$= \sum exp\left[-\frac{1}{2}(\underline{x}(k) - \underline{\hat{x}}_+)^T Q^{-1}(\underline{x}(k) - \underline{\hat{x}}_+)\right]$$

$$- \sum exp\left[-\frac{1}{2}(\underline{x}(k) - \underline{\hat{x}}_-)^T Q^{-1}(\underline{x}(k) - \underline{\hat{x}}_-)\right] \quad (12.27)$$

where the first (second) sum is over all the points $\underline{\hat{x}}_+ \in P_{n,d}(1)$ ($\underline{\hat{x}}_- \in P_{n,d}(-1)$). Now consider the nonlinear channel

$$\hat{x}(k) = s(k) + 0.5s(k-1) - 0.9[s(k) + 0.5s(k-1)]^3 \quad (12.28)$$

and white Gaussian noise $e(k)$ with $E[e^2(k)] = 0.2$. For this case, the optimal decision region for $n = 2$ and $d = 0$,

$$[\underline{x}(k) \in R^2 | p_1[\underline{x}(k)|\underline{\hat{x}}(k) \in P_{2,0}(1)] - p_{-1}[\underline{x}(k)|\underline{\hat{x}}(k) \in P_{2,0}(-1)] \geq 0] \quad (12.29)$$

is shown in Figure 12.2 as the shaded area. The elements of the sets $P_{2,0}(1)$ and $P_{2,0}(-1)$ are illustrated in Figure 12.2 by the "o" and "*", respectively. From Figure 12.2 we see that the optimal decision boundary for this case is severely nonlinear. We now use the RLS and LMS fuzzy adaptive filters to solve this specific equalization problem (channel (12.28), $e(k)$ white Gaussian with variance

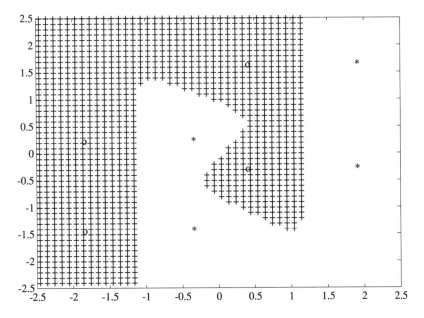

Figure 12.2 Optimal decision region for the channel (12.28), Gaussian white noise with variance $\sigma_e^2 = 0.2$, and equalizer order $n = 2$ and lag $d = 0$, where the horizontal-axis denotes $x(k)$ and the vertical-axis denotes $x(k-1)$.

0.2, equalizer order $n = 2$ and lag $d = 0$) under various conditions (Examples 12.1–12.4).

Example 12.1. Here, we used the RLS fuzzy adaptive filter without any linguistic information. We chose $\lambda = 0.999, \sigma = 0.1$, $m_1 = m_2 = 9$, and $\mu_{F_i^j}(x_i) = exp[-\frac{1}{2}(\frac{x_i - \bar{x}_i^j}{0.3})^2]$ with $\bar{x}_i^j = -2, -1.5, -1, -0.5, 0, 0.5, 1, 1.5, 2$ for $j = 1, 2, \ldots, 9$, respectively, where $i = 1, 2$, $x_1 = x(k)$ and $x_2 = x(k-1)$. For the same realization of the sequence $s(k)$ and the same randomly chosen initial parameters $\underline{\theta}(0)$ (within $[-0.3, 0.3]$), we simulated three cases: the adaptation (12.9)–(12.11) stopped at (1) $k = 30$, (2) $k = 50$, and (3) $k = 100$. The final decision regions, $[\underline{x}(k) \in R^2 | f_k(\underline{x}(k)) \geq 0]$ for the three cases are shown in Figures 12.3–12.5, respectively as shaded area. From Figures 12.3–12.5 we see that the decision regions obtained from the RLS fuzzy adaptive filter tended to converge toward the optimal decision region.

Example 12.2. Next, we used the RLS fuzzy adaptive filter incorporating the following linguistic information about the decision region. From the geometric formulation we see that the equalization problem is equivalent to determining a decision boundary in the input space of the equalizer. Suppose that there are human experts who are very familiar with the specific situation, such that although they cannot draw the specific decision boundary in the input space of the equalizer,

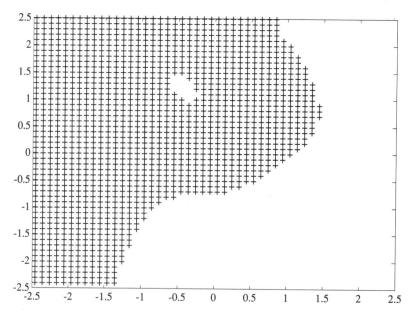

Figure 12.3 Decision region of the RLS fuzzy adaptive filter without using any linguistic information and when the adaptation stopped at $k = 30$, where the horizontal-axis denotes $x(k)$ and the vertical-axis denotes $x(k-1)$.

Sec. 12.4 Application to Nonlinear Channel Equalization

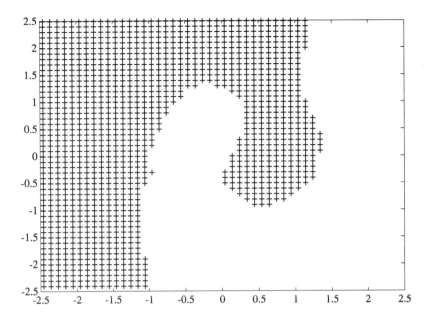

Figure 12.4 Decision region of the RLS fuzzy adaptive filter without using any linguistic information and when the adaptation stopped at $k = 50$, where the horizontal-axis denotes $x(k)$ and the vertical-axis denotex $x(k-1)$.

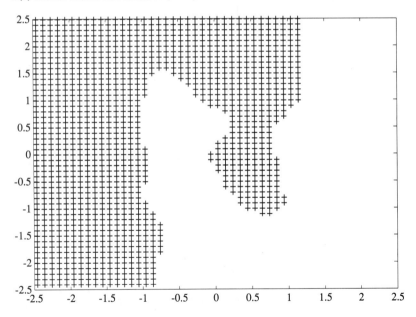

Figure 12.5 Decision region of the RLS fuzzy adaptive filter without using any linguistic information and when the adaptation stopped at $k = 100$, where the horizontal-axis denotes $x(k)$ and the vertical-axis denotex $x(k-1)$.

they can assign degrees to different regions in the input space which reflect their belief that the regions should belong to 1-catalog or -1-catalog. Take Figure 12.2 as an example. We see from Figure 12.2 that the difficulty is to determine which catalog the middle portion should belong to. In other words, as we move away from the middle portion, we have less and less uncertainty about which catalog the region should belong to. For example, for the left-most region in Figure 12.2, we have more confidence that it should belong to the 1-catalog rather than the -1-catalog. Similarly, for the right-most region in Figure 12.2, we have more confidence that it should belong to the -1-catalog rather than the 1-catalog. Also, we assume that the human experts know that a portion of the boundary is somewhere around $x(k) = -1.2$ for $x(k-1)$ less than 1 and around $x(k) = 1.2$ for $x(k-1)$ greater than -1. To make these observations specific, we have the fuzzy rules shown in Figure 12.6 where the membership functions N3, N2, and so on are the $\mu_{F_i^j}$'s defined in Example 12.1. We have 48 rules in Figure 12.6, corresponding to the boxes with numbers; for example, the bottom-left box corresponds to the rule: "IF $x(k)$ is N4 and $x(k-1)$ is N4, THEN f_k is G," where f_k is the filter output,

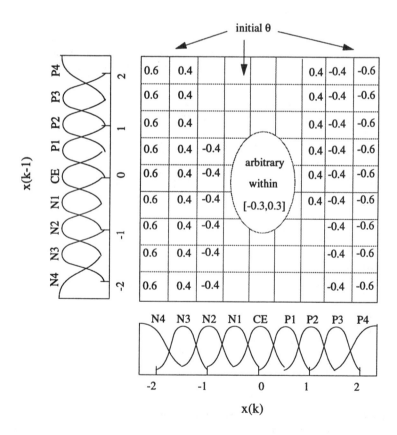

Figure 12.6 Illustration of some fuzzy rules about the decision region.

Sec. 12.4 Application to Nonlinear Channel Equalization

and the center of μ_G is 0.6. Because the filter output f_k is a weighted average of these centers (see (12.3)), the numbers 0.6, 0.4, −0.4, −0.6 in Figure 12.6 reflect our belief that the regions should correspond to the 1-catalog or the -1-catalog. For example, if the input point $[x(k), x(k-1)]$ falls in the left-most region of Figure 12.6, then we have more confidence that the transmitted $s(k)$ should be 1 rather than -1, and we represent this confidence by assigning the center of the fuzzy term in the corresponding THEN part to be 0.6.

It should be emphasized that the rules in Figure 12.6 provide very fuzzy information about the decision region because (1) the regions are fuzzy, that is, there are no clear boundaries between the regions, and (2) the numbers 0.6, 0.4, -0.4, -0.6 are conservative; that is, they are away from the real transmitted values 1 or -1. We now show that although these rules are fuzzy, the adaptation speed is greatly improved by incorporating them into the RLS fuzzy adaptive equalizer (filter). Figure 12.7 shows the final decision region determined by the RLS fuzzy adaptive filter, $[\underline{x}(k) \in R^2 | f_k(\underline{x}(k)) \geq 0]$ (shaded area), when the adaptation stopped at $k = 30$ after the rules in Figure 12.6 were incorporated, where the $\mu_{F_i^j}$'s and the sequence $s(k)$ were the same as those in Example 6.1. Comparing Figures 12.7 and 12.3, we see that the adaptation speed was greatly improved by incorporating these fuzzy rules.

Example 12.3. Here we used the LMS fuzzy adaptive filter without any linguistic information. We chose: $M = 20, \alpha = 0.05$, the initial $\theta^l(0)$ randomly

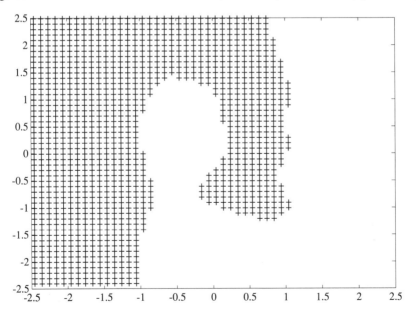

Figure 12.7 Decision region of the RLS fuzzy adaptive filter after incorporating the fuzzy rules illustrated in Figure 12.6 and when the adaptation stopped at $k = 30$, where the horizontal-axis denotes $x(k)$ and the vertical-axis denotes $x(k-1)$.

in $[-0.3, 0.3]$, $\bar{x}_i^j(0)$'s randomly in $[-2, 2]$, and $\bar{\sigma}_i^j(0)$'s randomly in $[0.1, 0.3]$. For the same sequence $s(k)$ (in Example 12.1) and the same initial parameters, we simulated the cases when the adaptation algorithm (12.16)–(12.18) stopped at: (1) $k = 100$, (2) $k = 200$, and (3) $k = 500$. The decision regions for the three cases are shown in Figures 12.8–12.10, respectively.

Example 12.4. Next, we used the LMS fuzzy adaptive filter and incorporated some of the fuzzy rules in Figure 12.6. We still chose $M = 20$ and $\alpha = 0.05$ and used the same $s(k)$ sequence. Since the filter is constructed from 20 rules, whereas Figure 12.6 contains 48 rules, we can only chose a portion of the rules in Figure 12.6 to construct the initial LMS fuzzy adaptive filter. We chose 20 rules arbitrarily from the boxes labeled 0.4 and -0.4. The final decision region for this case, when the adaptation stopped at $k = 100$, is shown in Figure 12.11. Comparing Figures 12.11 and 12.8, we see that the adaptation speed was improved by incorporating these fuzzy rules.

Example 12.5. Here, we considered the same situation as in Example 12.1, except that we chose $d = 1$ rather than $d = 0$. The optimal decision region for this case is shown in Figure 12.12. Figures 12.13 and 12.14 show the final decision regions determined by the RLS fuzzy adaptive filter when the adaptation stopped at $k = 20$ and $k = 50$, respectively. We also simulated the LMS fuzzy adaptive filter for this case, and the final decision region was similar to Figure 12.14 when

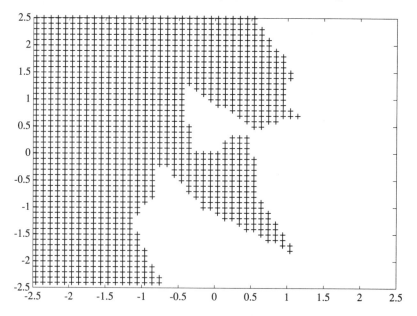

Figure 12.8 Decision region of the LMS fuzzy adaptive filter without using any linguistic information and when the adaptation stopped at $k = 100$, where the horizontal-axis denotes $x(k)$ and the vertical-axis denotes $x(k - 1)$.

Sec. 12.4　Application to Nonlinear Channel Equalization

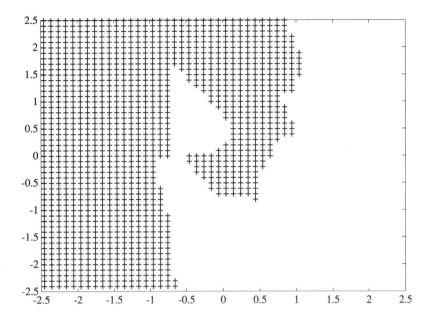

Figure 12.9　Decision region of the LMS fuzzy adaptive filter without using any linguistic information and when the adaptation stopped at $k = 200$, where the horizontal-axis denotes $x(k)$ and the vertical-axis denotes $x(k-1)$.

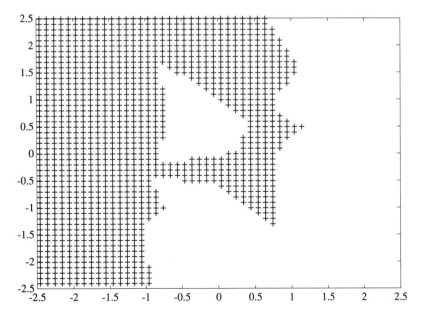

Figure 12.10　Decision region of the LMS fuzzy adaptive filter without using any linguistic information and when the adaptation stopped at $k = 500$, where the horizontal-axis denotes $x(k)$ and the vertical-axis denotes $x(k-1)$.

198 Fuzzy Adaptive Filters   Chap. 12

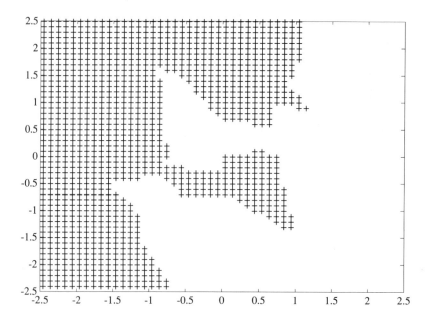

Figure 12.11 Decision region of the LMS fuzzy adaptive filter after incorporating some of the fuzzy rules illustrated in Figure 12.6 and when the adaptation stopped at $k = 100$, where the horizontal-axis denotes $x(k)$ and the vertical-axis denotes $x(k-1)$.

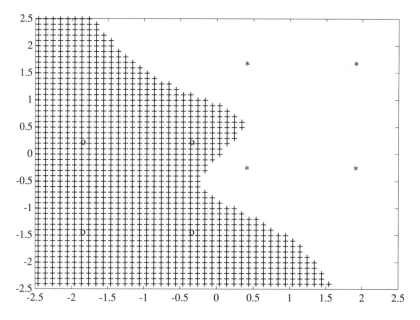

Figure 12.12 Optimal decision region for the channel (12.28), Gaussian white noise with variance $\sigma_e^2 = 0.2$, and equalizer order $n = 2$ and lag $d = 1$, where the horizontal-axis denotes $x(k)$ and the vertical-axis denotes $x(k-1)$.

Sec. 12.4　Application to Nonlinear Channel Equalization

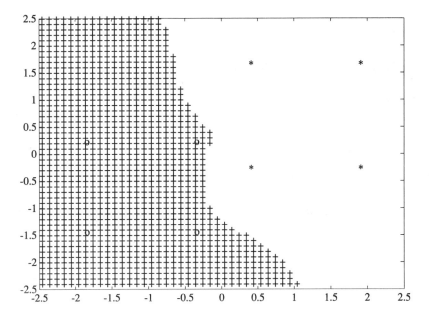

Figure 12.13 Decision region of the RLS fuzzy adaptive filter for the case of Example 12.5 when the adaptation stopped at $k = 20$, where the horizontal-axis denotes $x(k)$ and the vertical-axis denotes $x(k-1)$.

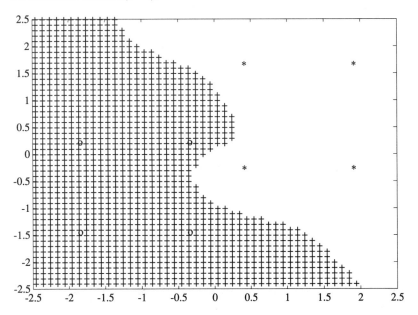

Figure 12.14 Decision region of the RLS fuzzy adaptive filter for the case of Example 12.5 when the adaptation stopped at $k = 50$, where the horizontal-axis denotes $x(k)$ and the vertical-axis denotes $x(k-1)$.

the adaptation stopped at $k = 300$. Since we showed this kind of comparison in Examples 12.1–12.4, we omit the details.

Example 12.6. In this final example, we compare the bit error rates achieved by the optimal equalizer (12.25) and the fuzzy adaptive equalizers for different signal-to-noise ratios, for the channel (12.28) with equalizer order $n = 2$ and lag $d = 1$. The optimal bit error rate was computed by applying the optimal equalizer (12.25) to a realization of 10^6 points of the sequences $s(k)$ and $e(k)$. For the RLS fuzzy adaptive filter, we chose the filter parameters to be the same as in Example 12.1. For the LMS fuzzy adaptive filter, the parameters were chosen as in Example 12.3. We ran the RLS and LMS fuzzy adaptive filters for the first 1,000 points in the same 10^6 point realization of $s(k)$ and $e(k)$ as for the optimal equalizer, and then used the trained fuzzy equalizers to compute the bit error rate for the same 10^6 point realization. Figure 12.15 shows the bit error rates of the optimal equalizer and the two fuzzy equalizers for different signal-to-noise ratios, where the bit error rate curves for the RLS and LMS fuzzy equalizers are indistinguishable. We see from Figure 12.15 that the bit error rates of the fuzzy equalizers are very close to the optimal one.

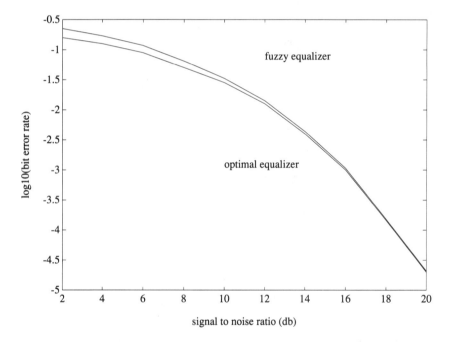

Figure 12.15 Comparison of bit error rates achieved by the optimal and fuzzy equalizers (Example 12.6).

12.5 CONCLUDING REMARKS

In this chapter, we developed two nonlinear adaptive filters based on adaptive fuzzy systems, namely, RLS and LMS fuzzy adaptive filters. The key elements of the fuzzy adaptive filters are a fuzzy logic system, which is constructed from a set of fuzzy IF-THEN rules, and an adaptive algorithm for updating the parameters in the fuzzy system. For the RLS fuzzy adaptive filter, an RLS type of algorithm is used, and for the LMS fuzzy adaptive filter, an LMS type of algorithm is used. The most important advantage of the fuzzy adaptive filters is that linguistic information from human experts (in the form of fuzzy IF-THEN rules) can be directly incorporated into the filters. If no linguistic information is available, the fuzzy adaptive filters become well-defined nonlinear adaptive filters, similar to the polynomial, neural nets, or radial basis function adaptive filters. We applied the two fuzzy adaptive filters to nonlinear channel equalization problems. Simulation results showed that (1) the fuzzy adaptive filters worked quite well without using any linguistic information; (2) by incorporating some linguistic rules into the fuzzy adaptive filters, the adaptation speed was greatly improved; and (3) the bit error rates of the fuzzy equalizers were close to that of the optimal equalizer.

13
CONCLUSIONS

13.1 GENERAL CONCLUSIONS

Fuzzy control is a controversial field. An emotional resistance to fuzzy control is due to the exaggerated claims made by some fuzzy researchers. From Chapter 2 we see that fuzzy controllers are not some magical controllers, they are simply *rule-based controllers*. Conventional control starts with a mathematical model of the system, and controllers are designed for the model. Fuzzy control starts with heuristics and human expertise, and controllers are designed by synthesizing these heuristics and human expertise. See Figure 13.1. Although the starting points of the two approaches are different, the end products are the same—nonlinear controllers for nonlinear systems. Therefore, fuzzy control theory should be viewed as a subset of nonlinear control theory in which the nonlinear controllers have a special structure.

What is this structure? In Chapter 2 we presented a systematic description of fuzzy logic systems which, when used as controllers, are the fuzzy controllers in the literature. Although fuzzy logic is a broad field which comprises many concepts and principles, the concepts and principles which have been really used in fuzzy logic systems are not so many. Chapter 2 summarized these concepts and principles, and a careful study of the chapter should be sufficient for understanding the essence of fuzzy logic systems.

Because of the controversy around fuzzy logic, the field is underdeveloped and leaves many fundamental questions open. Two of the questions are: how to determine the membership functions, and how to guarantee the stability of fuzzy control systems. This book addresses these two problems. In Chapters 3 to 6, we proposed four different methods for determining the membership functions.

Sec. 13.1 General Conclusions

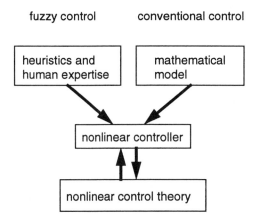

Figure 13.1 Fuzzy control is a subset of nonlinear control.

Chapter 3 proposed a gradient descent training algorithm for tuning the parameters of the membership functions based on input-output pairs. Chapter 4 determined the fuzzy logic systems by selecting the significant fuzzy basis functions. Chapter 5 proposed a simple one-pass procedure for designing fuzzy logic systems based on the table-lookup illustration of a fuzzy rule base. And Chapter 6 constructed the fuzzy logic systems by combining the optimal fuzzy logic system with clustering. All these methods were simulated for a number of control and signal processing problems, and the reader should get a feeling of the performance of these methods by observing these simulation results.

In Chapters 8 and 9 we developed a number of stable adaptive fuzzy controllers which are classified according to whether they can incorporate fuzzy control rules or fuzzy descriptions of the systems and whether the fuzzy logic systems involved are linearly or nonlinearly parameterized. We achieved stability by appending the adaptive fuzzy controller with a supervisory controller which pushed the state vector back to the stable region when the state vector tended to diverge. An advantage of this approach is that there is much freedom in designing the fuzzy controller because we need not worry about stability when we design the fuzzy controller—the supervisory controller takes care of the stability requirement. As a result, we may be able to design a high-performance fuzzy controller. Chapters 10 and 11 showed the extensions of the approach to more general nonlinear systems and to identification problems. Chapter 12 showed that adaptive fuzzy systems can be used as equalizers for nonlinear communication channels.

In summary, this book proposed a number of approaches to the two fundamental problems of fuzzy systems: how to determine the membership functions, and how to guarantee the stability of fuzzy control systems. This is only the beginning of developing a comprehensive theory of fuzzy systems and fuzzy control; much work remains to be done.

13.2 FUTURE RESEARCH

It is difficult to make an accurate prediction of future research because new discoveries enhance our knowledge base and change our original understanding. This is especially true for a young researcher who entered this fuzzy field less than three years ago and for a field which is underdeveloped due to strong criticism—let the reader be advised.

An In-depth Analysis of Fuzzy Logic Systems

One contribution of fuzzy logic is that it provides a systematic procedure of transforming a knowledge base into a nonlinear mapping, where the knowledge base consists of a collection of fuzzy IF-THEN rules (see Chapter 2). When this nonlinear mapping is used as a controller, it is called the fuzzy controller in the literature. Because we do not need the search operations as in the conventional expert systems, fuzzy logic systems—a particular kind of rule-based systems—are especially suitable for real-time applications like on-line control and decision making. In the transformation procedure from knowledge base to nonlinear mapping, some information must be lost, but what does this "information" mean? Is the "fuzzy entropy" in the literature [Klir and Folger, 27] a good measure of this information? We need a rigorous theory to guide this transformation. How is such a theory developed?

There may be two approaches. In the first approach, we set up some criterion which measures the "goodness" of the transformation from knowledge base to nonlinear mapping. Then the operations in inference engine, fuzzifier, and defuzzifier are chosen in such a way that this criterion is optimized. This criterion may be some kind of an "information measure." The key in this approach is to choose the criterion which should make sense. Also, the structures and operations in the blocks of the fuzzy logic system should be generalized such that we have a larger space to search for the optimal system and that the optimization problem is easier to solve. In the second approach, we put the fuzzy logic system into the particular problem and optimize the structure and operations of the fuzzy logic system based on a criterion which makes clear sense for the whole system. In this way we bypass the difficulty of choosing an "absolute" information measure required in the first approach. We also need to generalize the structure and operations of the fuzzy logic system. The limitation of this approach is that the insight gained is only applicable to the particular problem.

A Rigorous Theory of Fuzzy Control

From Figure. 13.1 we see that fuzzy control is a subset of nonlinear control; therefore, fuzzy control should be developed as a branch of nonlinear control theory. Obviously, we have not had this branch yet. Because the difference between fuzzy control and conventional nonlinear control is that fuzzy controllers have a particular structure, existing nonlinear control theory which is applicable to this particular structure can be directly used to analyze fuzzy controllers. However, we know that conventional nonlinear control theory is not well developed and many fundamental questions remain open. Therefore, the development of fuzzy control theory should parallel the development of conventional nonlinear control theory. We may develop a rigorous theory of fuzzy control along the following directions:

- *Robustness analysis of fuzzy controllers.* Many papers of fuzzy control claim that fuzzy controllers are robust with respect to variation in the model and in working conditions. However, this claim is based on simulations, not on theoretical analysis. A control theorist may immediately dismiss this claim and view this as one of the exaggerated claims of fuzzy control. However, as in many fields of science and engineering, simulations and experiments discover new phenomena and a detailed analysis of them may establish new theories. From a conceptual point of view, fuzzy controllers may indeed be robust because they are constructed from heuristics and human expertise, not from mathematical models. Therefore, the inaccuracy in the models should have less influence on the controller. An in-depth analysis of the robustness of fuzzy controllers is needed.
- *Stability analysis of nonadaptive fuzzy controllers.* There are some studies of this kind (for example, Langari and Tomizuka [29]), but the results were strong sufficient conditions which greatly restrict the design flexibility of fuzzy controllers. We need a better theory.
- *Systematic design procedure of fuzzy controllers.* Based on the preceding robustness and stability analyses, systematic design procedures of fuzzy controllers should be developed.
- *Analysis and design of new adaptive fuzzy controllers.* Many of the recent developments in the nonlinear adaptive control literature should be very useful.

Toward a Theory of Intelligent Control

The starting point of conventional control is a differential (or difference) equation, say $\dot{x} = f(x) + g(x)u$. Then a huge theory is developed to show how to control the systems represented by the differential equation. In practice, however, we are facing large-scale hybrid systems which combine control, communication, supervision, and human operators together. Differential equations derived from physical laws are clearly not sufficient to formulate such systems. At present the design of such hybrid control systems is ad hoc in nature. We need a theory to analyze such hybrid control systems and to guide our designs; such theory is usually called intelligent control theory. There are two main approaches to intelligent control: one that combines differential equations with expert systems in AI (artificial intelligence) the so-called expert control, and the other that combines differential equations with discrete event systems or Markov chain. The first approach is practically useful but is difficult to analyze because the formulations of differential equations and expert systems are fundamentally different—one is based on mathematical formulas and the other is based on symbolic AI. The second approach is mathematically well developed but the theory is complex and applications to real practical problems are not easy. Because of the complexity and diversity of large-scale hybrid systems, these two approaches should not be the only valid approaches. We think that fuzzy logic systems are very useful in helping to establish an intelligent control theory. Before we show examples of how to use fuzzy logic systems in the mathematical formulation of intelligent control systems, let us first make it clear what we mean by an intelligent control system and an intelligent control theory.

In our approach, intelligent control systems are defined as control systems which include low-level feedback control and high-level supervision and planning. See Figure 13.2. An intelligent control theory is a theory on the analysis and design of such hierarchically organized control systems. The low-level process is usually represented by differential equations. The question here is how to formulate the high-level supervision and planning. We think that the operations of the high-level supervision and planning can be represented by a set of fuzzy and nonfuzzy IF-THEN rules because in many existing systems these operations are performed by humans and IF-THEN rules are the most straightforward framework to represent these human operations. Because fuzzy logic systems are rule-based systems and nonfuzzy IF-THEN rules are special cases of fuzzy IF-THEN rules, it should be a valid approach to represent the supervision and planning levels by fuzzy logic systems. We call this approach the differential equation plus fuzzy logic system approach. As compared with the differential equation plus expert system approach, the advantage of this approach is that all levels are represented by mathematical formulas (recall that fuzzy logic systems transform a rule base into a nonlinear mapping). Therefore, mathematical systems theory can be used to analyze such systems in a rigorous way. As compared with the differential equation plus discrete event system approach, the advantage of this approach is that IF-THEN rules are a more powerful framework to represent high-level operations than discrete event systems or Markov chain because human knowledge, which is more important at

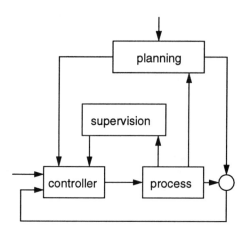

Figure 13.2 A general architecture of intelligent control systems.

the high levels than at the low levels, can be directly incorporated into the IF-THEN rule framework, while incorporating human knowledge into discrete event systems or Markov chain is not straightforward. Next, we show very briefly some of the preliminary ideas of how to use this approach to formulate a robot control system and the intelligent vehicle/highway systems.

Consider the situation in Figure 13.3, where our task is to control the robot from point A to point B without hitting the obstacles. Suppose we have the knowledge that the robot should turn left at intersections C and D and turn right at intersection E. We now use the three-level architecture of Figure 13.2 to formulate the whole control system: the process represents the dynamics of the robot; the low-level controller is a conventional tracking controller; the supervision level consists of some safeguard control rules which "overwrite" the tracking controller in extreme situations; and the planning level determines the desired trajectory the robot should follow. More specifically, the supervision level may contain the following two rules (where $d_1(d_2)$ is the distance from the robot to the left (right) obstacle):

IF d_1 is much larger than d_2, THEN move the robot to right. (13.1)

IF d_1 is much smaller than d_2, THEN move the robot to left. (13.2)

Similarly, the planning level may contain the following rules:

IF d_1 is larger than d_2,

THEN adjust the desired trajectory to the right slightly. (13.3)

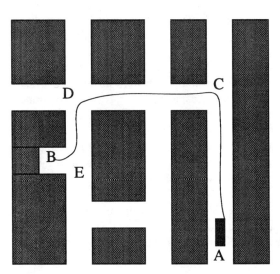

Figure 13.3 Control the robot from point A to point B without hitting the obstacles.

> IF d_1 is smaller than d_2,
>
> THEN adjust the desired trajectory to the left slightly. (13.4)
>
> IF the distance to point C is small,
>
> THEN adjust the desired trajectory to the left greatly. (13.5)
>
> IF the distance to point D is small,
>
> THEN adjust the desired trajectory to the left greatly. (13.6)
>
> IF the distance to point E is small,
>
> THEN adjust the desired trajectory to the right greatly. (13.7)

More rules are needed in real implementation; for example, we may need more rules at the intersections. Using these rules as the knowledge bases of fuzzy logic systems, we obtain an entire control system which is represented by the following equations:

$$\dot{x} = f(x, u) \qquad (13.8)$$

$$u = u_t + u_s \qquad (13.9)$$

$$\dot{c} = g(c, r) \qquad (13.10)$$

where (13.8) represents the dynamics of the robot, (13.9) shows that the controller comprises two parts: a conventional tracking controller u_t and a supervisory controller u_s which is a fuzzy logic system constructed from the supervisional rules like (13.1) and (13.2), and (13.10) shows the desired trajectory c with g being a fuzzy logic system constructed from the rules like (13.3)–(13.7) and r as measurements like d_1 and d_2. This is a mathematical formulation of the hierarchically organized control system—a first step toward developing a theory for it.

Next we show that the same idea can be used to formulate the intelligent vehicle/highway systems. One way to improve the efficiency of highway systems is to group cars into platoons; see Figure 13.4. The cars in the same platoon have a very short distance between each other and travel at the same speed. The tasks here are: (1) guarantee the safety of the system, (2) determine the desired speeds of the platoons such that the freeway achieves its maximum efficiency, and (3) control the platoons to the desired speeds. We now use the three-level architecture of Figure 13.2 to formulate this control system: the process represents the dynamics of the cars; the low-level controller represents lateral and longitudinal controllers; the supervision level comprises safeguard controllers which are constructed from rules like

> IF the distance to left lane boundary is short,
>
> THEN move the platoon to right. (13.11)

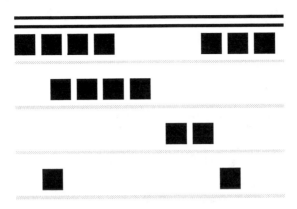

Figure 13.4 A way to improve freeway efficiency is to group cars into platoons.

for lateral control; and rules like

> *IF the speed of a car in the platoon is reduced greatly,*
>
> *THEN control the whole platoon to the speed of this car.* (13.12)

for longitudinal control; and the planning level is a decision-making system which contains rules like

> *IF the distance to the platoon ahead is long,*
>
> *THEN increase the desired speed of the platoon.* (13.13)
>
> *IF the density of cars in the next section is high,*
>
> *THEN reduce the desired speed of the platoon.* (13.14)

Of course, more rules are needed in real implementation. The mathematical formulation of the whole system is the same as (13.8)–(13.10).

From the preceding two examples, we see that equations (13.8)–(13.10) are a good mathematical formulation of the intelligent control systems of Figure 13.2. The special characteristic of this formulation is that u_s and g are fuzzy logic systems. If this can be justified as a general and valid formulation, then a whole set of theories—analysis, design, adaptation, and so on—should be developed for it.

Appendix

PROOFS OF THE UNIVERSAL APPROXIMATION AND STABILITY THEOREMS

We use the following Stone-Weierstrass Theorem to prove the Universal Approximation Theorem in Chapter 2.

Stone-Weierstrass Theorem [Rudin, 60].   Let Z be a set of real continuous functions on a compact set U. If: (1) Z is an *algebra*, that is, the set Z is closed under addition, multiplication, and scalar multiplication; (2) Z *separates points on* U, that is, for every $\underline{x}, \underline{y} \in U, \underline{x} \neq \underline{y}$, there exists $f \in Z$ such that $f(\underline{x}) \neq f(\underline{y})$; and (3) Z *vanishes at no point of* U, that is, for each $\underline{x} \in U$ there exists $f \in Z$ such that $f(\underline{x}) \neq 0$; then the uniform closure of Z consists of all real continuous functions on U, that is, (Z, d_∞) is dense in $(C[U], d_\infty)$.

Let Y be the set of all fuzzy logic systems in the form of (2.46). In order to use the Stone-Weierstrass Theorem to prove the Universal Approximation Theorem, we need to show that Y is an algebra, Y separates points on U, and Y vanishes at no point of U. The following Lemmas A.1–A.3 prove that Y has these properties.

Lemma A.1. (Y, d_∞) is an algebra.

Proof. Let $f_1, f_2 \in Y$, so that we can write them as

$$f_1(\underline{x}) = \frac{\sum_{j=1}^{K1} \left(\bar{z}1^j \prod_{i=1}^{n} \mu_{A1_i^j}(x_i) \right)}{\sum_{j=1}^{K1} \left(\prod_{i=1}^{n} \mu_{A1_i^j}(x_i) \right)} \qquad (A.1)$$

$$f_2(\underline{x}) = \frac{\sum_{j=1}^{K2} \left(\bar{z}2^j \prod_{i=1}^{n} \mu_{A2_i^j}(x_i) \right)}{\sum_{j=1}^{K2} \left(\prod_{i=1}^{n} \mu_{A2_i^j}(x_i) \right)} \qquad (A.2)$$

Hence,

$$f_1(\underline{x}) + f_2(\underline{x}) = \frac{\sum_{j1=1}^{K1} \sum_{j2=1}^{K2} \left(\bar{z1}^{j1} + \bar{z2}^{j2}\right) \left(\Pi_{i=1}^n \mu_{A1_i^{j1}}(x_i) \mu_{A2_i^{j2}}(x_i)\right)}{\sum_{j1=1}^{K1} \sum_{j2=1}^{K2} \left(\Pi_{i=1}^n \mu_{A1_i^{j1}}(x_i) \mu_{A2_i^{j2}}(x_i)\right)} \quad (A.3)$$

Since $\mu_{A1_i^{j1}}$ and $\mu_{A2_i^{j2}}$ are Gaussian in form, their product $\mu_{A1_i^{j1}} \mu_{A2_i^{j2}}$ is also Gaussian in form (this can be verified by straightforward algebraic operations). Hence, (A.3) is the same form as (2.46), so that $f_1 + f_2 \in Y$. Similarly,

$$f_1(\underline{x}) f_2(\underline{x}) = \frac{\sum_{j1=1}^{K1} \sum_{j2=1}^{K2} \left(\bar{z1}^{j1} \bar{z2}^{j2}\right) \left(\Pi_{i=1}^n \mu_{A1_i^{j1}}(x_i) \mu_{A2_i^{j2}}(x_i)\right)}{\sum_{j1=1}^{K1} \sum_{j2=1}^{K2} \left(\Pi_{i=1}^n \mu_{A1_i^{j1}}(x_i) \mu_{A2_i^{j2}}(x_i)\right)} \quad (A.4)$$

which is also in the same form of (2.46). Hence, $f_1 f_2 \in Y$. Finally, for arbitrary $c \in R$,

$$cf_1(\underline{x}) = \frac{\sum_{j=1}^{K1} (c\bar{1}^j)(\Pi_{i=1}^n \mu_{A1_i^j}(x_i))}{\sum_{j=1}^{K1} (\Pi_{i=1}^n \mu_{A1_i^j}(x_i))} \quad (A.5)$$

which is again in the form of (2.46). Hence, $cf_1 \in Y$.   Q.E.D.

Lemma A.2. (Y, d_∞) separates points on U.

Proof. We prove this by constructing a required f. That is, we specify the number of fuzzy sets defined in U and R, the parameters of the Gaussian membership functions, the number of fuzzy rules, and the statements of fuzzy rules, such that the resulting f (in the form of (2.46)) has the property that $f(\underline{x}^0) \neq f(\underline{y}^0)$ for arbitrarily given $\underline{x}^0, \underline{y}^0 \in U$ with $\underline{x}^0 \neq \underline{y}^0$. Let $\underline{x}^0 = (x_1^0, x_2^0, \ldots, x_n^0)$ and $\underline{y}^0 = (y_1^0, y_2^0, \ldots, y_n^0)$. If $x_i^0 \neq y_i^0$, we define two fuzzy sets, $(A_i^1, \mu_{A_i^1})$ and $(A_i^2, \mu_{A_i^2})$, in the ith subspace of U, with

$$\mu_{A_i^1}(x_i) = exp\left[-\frac{(x_i - x_i^0)^2}{2}\right] \quad (A.6)$$

$$\mu_{A_i^2}(x_i) = exp\left[-\frac{(x_i - y_i^0)^2}{2}\right] \quad (A.7)$$

If $x_i^0 = y_i^0$, then $A_i^1 = A_i^2$ and $\mu_{A_i^1} = \mu_{A_i^2}$. That is, only one fuzzy set is defined in the ith subspace of U. We define two fuzzy sets, (B^1, μ_{B^1}) and (B^2, μ_{B^2}), in the output universe of discourse R, with

$$\mu_{B^j}(z) = exp\left[-\frac{(z - \bar{z}^j)^2}{2}\right] \quad (A.8)$$

where $j = 1, 2$, and $\bar{z}^j$ will be specified later. We choose two fuzzy rules for the fuzzy rule base (that is, $M = 2$). Now we have specified all the design parameters except $\bar{z}^j$ ($j = 1, 2$). That is, we have already obtained a function f which is in the form of (2.46) with $M = 2$. With this f, we have

$$f(\underline{x}^0) = \frac{\bar{z}^1 + \bar{z}^2 \Pi_{i=1}^n exp\left[-(x_i^0 - y_i^0)^2/2\right]}{1 + \Pi_{i=1}^n exp\left[-(x_i^0 - y_i^0)^2/2\right]} = \alpha \bar{z}^1 + (1 - \alpha)\bar{z}^2 \quad \text{(A.9)}$$

$$f(\underline{y}^0) = \frac{\bar{z}^2 + \bar{z}^1 \Pi_{i=1}^n exp\left[-(x_i^0 - y_i^0)^2/2\right]}{1 + \Pi_{i=1}^n exp\left[-(x_i^0 - y_i^0)^2/2\right]} = \alpha \bar{z}^2 + (1 - \alpha)\bar{z}^1 \quad \text{(A.10)}$$

where

$$\alpha = \frac{1}{1 + \Pi_{i=1}^n exp\left[-(x_i^0 - y_i^0)^2/2\right]} \quad \text{(A.11)}$$

Since $\underline{x}^0 \neq \underline{y}^0$, there must be some i such that $x_i^0 \neq y_i^0$; hence, we have $\Pi_{i=1}^n exp[-(x_i^0 - y_i^0)^2/2] \neq 1$, or, $\alpha \neq 1 - \alpha$. If we choose $\bar{z}^1 = 0$ and $\bar{z}^2 = 1$, then $f(\underline{x}^0) = 1 - \alpha \neq \alpha = f(\underline{y}^0)$. Q.E.D.

Lemma A.3. (Y, d_∞) vanishes at no point of U.

Proof. By observing (2.46), we simply choose all $\bar{y}^l > 0$ ($l = 1, 2, \ldots, M$), that is, any $f \in Y$ with $\bar{y}^j > 0$ serves as the required f. Q.E.D.

Proof of the Universal Approximation Theorem. From (2.46), it is obvious that Y is a set of real continuous functions on U. The Universal Approximation Theorem is therefore a direct consequence of the Stone-Weierstrass Theorem and Lemmas A.1–A.3. Q.E.D.

Next, we prove the corollary of the Universal Approximation Theorem.

Proof of the Corollary. Since U is compact, $\int_U d\underline{x} = V < \infty$. Since continuous functions on U form a dense subset of $L_2(U)$ [Rudin, 60], for any $g \in L_2(U)$ there exists a continuous function $\bar{g}$ on U such that $(\int_U |g(\underline{x}) - \bar{g}(\underline{x})|^2 d\underline{x})^{1/2} < \epsilon/2$. By the Universal Approximation Theorem, there exists $f \in Y$ such that $sup_{\underline{x} \in U} |f(\underline{x}) - \bar{g}(\underline{x})| < \epsilon/(2V^{1/2})$. Hence, we have

$$\left(\int_U |f(\underline{x}) - g(\underline{x})|^2 d\underline{x}\right)^{1/2} \leq \left(\int_U |f(\underline{x}) - \bar{g}(\underline{x})|^2 d\underline{x}\right)^{1/2}$$

$$+ \left(\int_U |\bar{g}(\underline{x}) - g(\underline{x})|^2 d\underline{x}\right)^{1/2}$$

$$< \left(\int_U \left(sup_{\underline{x} \in U} |f(\underline{x}) - \bar{g}(\underline{x})|\right)^2 d\underline{x}\right)^{1/2} + \epsilon/2$$

$$< \left(\frac{\epsilon^2}{2^2 V} V\right)^{1/2} + \epsilon/2 = \epsilon \quad \text{(A.12)}$$

Q.E.D.

Proofs of the Universal Approximation and Stability Theorems 213

Proof of Theorem 8.1.

1. To prove $|\underline{\theta}_f| \leq M_f$, let $V_f = \frac{1}{2}\underline{\theta}_f^T \underline{\theta}_f$. If the first line of (8.42) is true, we have either $|\underline{\theta}_f| < M_f$ or $\dot{V}_f = -\gamma_1 \underline{e}^T P\underline{b}_c \underline{\theta}_f^T \underline{\xi}(\underline{x}) \leq 0$ when $|\underline{\theta}_f| = M_f$, that is, we always have $|\underline{\theta}_f| \leq M_f$. If the second line of (8.42) is true, we have $|\underline{\theta}_f| = M_f$ and $\dot{V}_f = -\gamma_1 \underline{e}^T P\underline{b}_c \underline{\theta}_f^T \underline{\xi}(\underline{x}) + \gamma_1 \underline{e}^T P\underline{b}_c \frac{|\underline{\theta}_f|^2 \underline{\theta}_f^T \underline{\xi}(\underline{x})}{|\underline{\theta}_f|^2} = 0$, that is, $|\underline{\theta}_f| \leq M_f$. Therefore, we have $|\underline{\theta}_f(t)| \leq M_f, \forall t \geq 0$. Using the same method, we can prove that $|\underline{\theta}_g(t)| \leq M_g, \forall t \geq 0$. From (8.44) we see that if $\theta_{gi} = \epsilon$, then $\dot{\theta}_{gi} \geq 0$; that is, we have $\theta_{gi} \geq \epsilon$ for all elements θ_{gi} of $\underline{\theta}_g$.

   In Section 8.2 we proved that $\bar{V}_e \leq \bar{V}$; therefore, $\frac{1}{2}\lambda_{min}|\underline{e}|^2 \leq \frac{1}{2}\underline{e}^T P\underline{e} \leq \bar{V}$; that is, $|\underline{e}| \leq (\frac{2\bar{V}}{\lambda_{min}})^{1/2}$. Since $\underline{e} = \underline{y}_m - \underline{x}$, we have $|\underline{x}| \leq |\underline{y}_m| + |\underline{e}| \leq |\underline{y}_m| + (\frac{2\bar{V}}{\lambda_{min}})^{1/2}$, which is (8.47). Finally, we prove (8.48). Since $\hat{f}(\underline{x}|\underline{\theta}_f)$ and $\hat{g}(\underline{x}|\underline{\theta}_g)$ are weighted averages of the elements of $\underline{\theta}_f$ and $\underline{\theta}_g$, respectively, we have $|\hat{f}(\underline{x}|\underline{\theta}_f)| \leq |\underline{\theta}_f| \leq M_f$ and $\hat{g}(\underline{x}|\underline{\theta}_g) \geq \epsilon$ (since all elements of $\underline{\theta}_g \geq \epsilon$). Therefore, from (8.8) we have

$$|u_c| \leq \frac{1}{\epsilon}\left(M_f + |y_m^{(n)}| + |\underline{k}|\left(\frac{2\bar{V}}{\lambda_{min}}\right)^{1/2}\right) \quad (A.13)$$

From (8.17) we have

$$|u_s| \leq \frac{1}{|g_L(\underline{x})|}\left[M_f + |f^U(\underline{x})| + (M_f + |g^U(\underline{x})|)|u_c|\right] \quad (A.14)$$

Combining (A.13) and (A.14) we obtain (8.48).

2. From (8.27) and (8.42)–(8.46), we have

$$\dot{V} = -\frac{1}{2}\underline{e}^T Q\underline{e} - g(\underline{x})\underline{e}^T P\underline{b}_c u_s + \underline{e}^T P\underline{b}_c w + I_1 \underline{e}^T P\underline{b}_c \frac{\phi_f^T \underline{\theta}_f \underline{\theta}_f^T \underline{\xi}(\underline{x})}{|\underline{\theta}_f|^2}$$

$$+ I_2 \underline{e}^T P\underline{b}_c \frac{\phi_{g+}^T \underline{\theta}_{g+} \underline{\theta}_{g+}^T \underline{\xi}_+(\underline{x})u_c}{|\underline{\theta}_{g+}|^2} + I_3 \phi_{g\epsilon}^T \underline{e}^T P\underline{b}_c \underline{\xi}_\epsilon(\underline{x})u_c \quad (A.15)$$

where $I_1 = 0(1)$ if the first (second) line of (8.42) is true, $I_2 = 0(1)$ if the first (second) line of (8.45) is true, $I_3 = 0(1)$ if the first (second) line of (8.44) is true. $\underline{\theta}_{g+}$ denotes the collection of θ_{gi}'s $> \epsilon$, $\underline{\theta}_{g\epsilon}$ denotes the collection of θ_{gi}'s $= \epsilon$, $\phi_{g+} = \underline{\theta}_{g+} - \underline{\theta}_{g+}^*$, $\phi_{g\epsilon} = \underline{\theta}_{g\epsilon} - \underline{\theta}_{g\epsilon}^*$, and $\underline{\xi}_+(\underline{x})$ ($\underline{\xi}_\epsilon(\underline{x})$) is the collection of the corresponding elements of $\underline{\xi}(\underline{x})$ with respect to $\underline{\theta}_{g+}$ ($\underline{\theta}_{g\epsilon}$). Now we show that the last three terms of (A.15) are nonpositive. First, the term with I_1. If $I_1 = 0$, the conclusion is trivial. Let $I_1 = 1$, which means that $|\underline{\theta}_f| = M_f$ and $\underline{e}^T P\underline{b}_c \underline{\theta}_f^T \underline{\xi}(\underline{x}) < 0$, we have $\phi_f^T \underline{\theta}_f = (\underline{\theta}_f - \underline{\theta}_f^*)^T \underline{\theta}_f = \frac{1}{2}[|\underline{\theta}_f|^2 - |\underline{\theta}_f^*|^2 + |\underline{\theta}_f - \underline{\theta}_f^*|^2] \geq 0$, since $|\underline{\theta}_f| = M_f \geq |\underline{\theta}_f^*|$. Therefore, the term with I_1 is nonpositive. Similarly, we can prove that the term with I_2 is nonpositive. Finally, from (8.44) and the fact that $\phi_{gi} = \theta_{gi} - \theta_{gi}^*$

$= \epsilon - \theta_{gi}^* \leq 0$, we have that the term with I_3 is also nonpositive. Therefore, we have

$$\dot{V} \leq -\frac{1}{2} \underline{e}^T Q \underline{e} - g(\underline{x}) \underline{e}^T P \underline{b}_c u_s + \underline{e}^T P \underline{b}_c w \qquad (A.16)$$

From (8.17) and $g(\underline{x}) > 0$, we have $g(\underline{x}) \underline{e}^T P \underline{b}_c u_s \geq 0$; therefore, (A.16) can be further simplified to

$$\dot{V} \leq -\frac{1}{2} \underline{e}^T Q \underline{e} + \underline{e}^T P \underline{b}_c w$$

$$\leq -\frac{\lambda_{Qmin} - 1}{2} |\underline{e}|^2 - \frac{1}{2} \left[|\underline{e}|^2 - 2\underline{e}^T P \underline{b}_c w + |P \underline{b}_c w|^2 \right] + \frac{1}{2} |P \underline{b}_c w|^2$$

$$\leq -\frac{\lambda_{Qmin} - 1}{2} |\underline{e}|^2 + \frac{1}{2} |P \underline{b}_c w|^2 \qquad (A.17)$$

where λ_{Qmin} is the minimum eigenvalue of Q. Integrating both sides of (A.17) and assuming that $\lambda_{Qmin} > 1$ (since Q is determined by the designer, we can choose such a Q), we have

$$\int_0^t |\underline{e}(\tau)|^2 d\tau \leq \frac{2}{\lambda_{Qmin} - 1} [|V(0)| + |V(t)|] + \frac{1}{\lambda_{Qmin} - 1} |P \underline{b}_c|^2 \int_0^t |w(\tau)|^2 d\tau$$
(A.18)

Define $a = \frac{2}{\lambda_{Qmin}-1}[|V(0)| + sup_{t \geq 0}|V(t)|]$ and $b = \frac{1}{\lambda_{Qmin}-1}|P\underline{b}_c|^2$, (A.18) becomes (8.49) (note that $sup_{t \geq 0}|V(t)|$ is finite because $\underline{e}, \underline{\phi}_f$ and $\underline{\phi}_g$ are all bounded).

3. If $w \in L_2$, then from (8.49) we have $\underline{e} \in L_2$. Because we have proven that all the variables in the right-hand side of (8.25) are bounded, we have $\underline{\dot{e}} \in L_\infty$. Using the Barbalat's Lemma [Slotine and Li, 66] (if $\underline{e} \in L_2 \cap L_\infty$ and $\underline{\dot{e}} \in L_\infty$, then $lim_{t \to \infty}|\underline{e}(t)| = 0$), we have $lim_{t \to \infty}|\underline{e}(t)| = 0$. Q.E.D.

Proof of Theorem 8.2. Comparing the error equation (8.25) of the first-type adaptive fuzzy controller with the error equation (8.33) of the second-type adaptive fuzzy controller, we see that they are the same if we replace w by v and $\underline{\xi}(\underline{x})$ by $\frac{\partial \hat{f}}{\partial \underline{\theta}_f}$ and $\frac{\partial \hat{g}}{\partial \underline{\theta}_g}$ (as in (8.33)). Also, as in (8.44), the adaptive law (8.58), (8.60), and (8.61) guarantees that $\sigma_{fi}^l \geq \sigma$, $\bar{y}_g^l \geq \epsilon$, and $\sigma_{gi}^l \geq \sigma$. Using exactly the same procedure as in the proof of Theorem 8.1, we can prove this theorem; therefore, we omit the details. Q.E.D.

Proof of Theorems 9.1 and 9.2. Using the same procedure as in the proof of Theorems 8.1 and 8.2, we can prove these theorems; therefore, we omit the details.

Proof of Theorem 11.1. From (2.46) we see that $\hat{g}$ and $\hat{h}$ are continuous functions and therefore satisfy the Lipschitz condition in the compact set K of (11.5).

That is, there exist constants b_g and b_h such that for all $(\underline{x}^{(1)}, \underline{u}), (\underline{x}^{(2)}, \underline{u}) \in K$

$$|\hat{g}(\underline{x}^{(1)}, \underline{u}|\Theta_g) - \hat{g}(\underline{x}^{(2)}, \underline{u}|\Theta_g)| \leq b_g |\underline{x}^{(1)} - \underline{x}^{(2)}| \quad (A.19)$$

$$|\hat{h}(\underline{x}^{(1)}, \underline{u}|\Theta_h) - \hat{h}(\underline{x}^{(2)}, \underline{u}|\Theta_h)| \leq b_h |\underline{x}^{(1)} - \underline{x}^{(2)}| \quad (A.20)$$

Define $\underline{e}_x \equiv \hat{\underline{x}} - \underline{x}$, then from (11.1) and (11.8) we have

$$\dot{\underline{e}}_x = A\underline{e}_x + \hat{g}(\hat{\underline{x}}, \underline{u}|\Theta_g^*) - g(\underline{x}, \underline{u}) \quad (A.21)$$

whose solution can be expressed as

$$\underline{e}_x(t) = \int_0^t e^{A(t-\tau)} [\hat{g}(\hat{\underline{x}}(\tau), \underline{u}(\tau)|\Theta_g^*) - g(\underline{x}(\tau), \underline{u}(\tau))]d\tau \quad (A.22)$$

Since A is a Hurwitz matrix, there exist positive constants c and α such that $\|e^{At}\| \leq ce^{-\alpha t}$ for all $t \geq 0$. Let $\epsilon_g = \frac{\epsilon\alpha}{2cb_h} e^{-\frac{cb_g}{\alpha}}$. Then from (A.22), (A.19), and (11.6) we have that

$$|\underline{e}_x(t)| \leq \int_0^t \|e^{A(t-\tau)}\| |\hat{g}(\hat{\underline{x}}(\tau), \underline{u}(\tau)|\Theta_g^*) - \hat{g}(\underline{x}(\tau), \underline{u}(\tau)|\Theta_g^*)|d\tau$$

$$+ \int_0^t \|e^{A(t-\tau)}\| |\hat{g}(\underline{x}(\tau), \underline{u}(\tau)|\Theta_g^*) - g(\underline{x}(\tau), \underline{u}(\tau))|d\tau$$

$$\leq \int_0^t ce^{-\alpha(t-\tau)} b_g |\underline{e}_x(\tau)|d\tau + \int_0^t ce^{-\alpha(t-\tau)} \frac{\epsilon\alpha}{2cb_h} e^{-\frac{cb_g}{\alpha}}d\tau$$

$$\leq cb_g \int_0^t e^{-\alpha(t-\tau)} |\underline{e}_x(\tau)|d\tau + \frac{\epsilon}{2b_h} e^{-\frac{cb_g}{\alpha}} \quad (A.23)$$

Using the Bellman-Gronwall Lemma [Hale, 17], we obtain

$$|\underline{e}_x(t)| \leq \frac{\epsilon}{2b_h} e^{-\frac{cb_g}{\alpha}} e^{cb_g \int_0^t e^{-\alpha(t-\tau)}d\tau}$$

$$\leq \frac{\epsilon}{2b_h} e^{-\frac{cb_g}{\alpha}} \left[1 + \int_0^t e^{-\alpha(t-\tau)}d\alpha(t-\tau)\right] \leq \frac{\epsilon}{2b_h} \quad (A.24)$$

Therefore, $|\hat{\underline{x}} - \underline{x}^0| \leq |\hat{\underline{x}} - \underline{x}| + |\underline{x} - \underline{x}^0| \leq \frac{\epsilon}{2b_h} + b$ from Assumption 11.1. Without loss of generality, we assume that $b_h \geq 1$; therefore, $(\hat{\underline{x}}, \underline{u}) \in K$. Hence, letting $\epsilon_h = \epsilon/2$, using (11.7), (A.20), and (A.24), and considering the fact that $(\underline{x}, \underline{u}) \in K$ and $(\hat{\underline{x}}, \underline{u}) \in K$, we obtain

$$|\hat{y}(t) - \underline{y}(t)| \leq |\hat{h}(\hat{\underline{x}}, \underline{u}|\Theta_h^*) - \hat{h}(\underline{x}, \underline{u}|\Theta_h^*)| + |\hat{h}(\underline{x}, \underline{u}|\Theta_h^*) - h(\underline{x}, \underline{u})|$$

$$\leq b_h |\underline{e}_x(t)| + \epsilon/2 \leq \epsilon \quad (A.25)$$

for all $t \in [0, T]$. \hfill Q.E.D.

Proof of Theorem 11.2.

1. From (11.26) and (11.28) we see that if $tr(\Theta_f \Theta_f^T) = M_f$, then, if $\underline{e}^T \Theta_f \underline{\xi}(x) \leq 0$, we have

$$\frac{d}{dt}[tr(\Theta_f \Theta_f^T)] = tr(\dot{\Theta}_f \Theta_f^T + \Theta_f \dot{\Theta}_f^T)$$

$$= tr(\gamma_1 \underline{e} \underline{\xi}^T(x) \Theta_f^T + \gamma_1 \Theta_f \underline{\xi}(x) \underline{e}^T)$$

$$= 2\gamma_1 \underline{e}^T \Theta_f \underline{\xi}(x) \leq 0 \quad (A.26)$$

where we use the properties of trace in the last equality; and if $\underline{e}^T \Theta_f \underline{\xi}(x) > 0$, we have

$$\frac{d}{dt}[tr(\Theta_f \Theta_f^T)] = 2tr\left(\gamma_1 \underline{e} \underline{\xi}^T(x) \Theta_f^T - \gamma_1 \frac{\underline{e}^T \Theta_f \underline{\xi}(x)}{tr(\Theta_f \Theta_f^T)} \Theta_f \Theta_f^T\right)$$

$$= 2\gamma_1 \underline{e}^T \Theta_f \underline{\xi}(x) - 2\gamma_1 \underline{e}^T \Theta_f \underline{\xi}(x) = 0 \quad (A.27)$$

Hence, we always have $tr(\Theta_f \Theta_f^T) \leq M_f$. Similarly, we can prove that $tr(\Theta_g \Theta_g^T) \leq M_g$. To prove $\hat{\underline{x}} \in L_\infty$, define the Lyapunov function candidate

$$V = \frac{1}{2}|\underline{e}|^2 + \frac{1}{2\gamma_1} tr(\Phi_f \Phi_f^T) + \frac{1}{2\gamma_2} tr(\Phi_g \Phi_g^T) \quad (A.28)$$

where $\Phi_f \equiv \Theta_f^* - \Theta_f$ and $\Phi_g \equiv \Theta_g^* - \Theta_g$. Subtracting (11.25) from (11.12) and using (11.23) and (11.24), we obtain the error dynamic equation

$$\dot{\underline{e}} = -\alpha \underline{e} + \Phi_f \underline{\xi}(x) + \Phi_g \underline{\xi}(x)u + \underline{w} \quad (A.29)$$

The time derivative of V along the solution of (A.29) is

$$\dot{V} = -\alpha|\underline{e}|^2 + \underline{e}^T \Phi_f \underline{\xi}(x) + \underline{e}^T \Phi_g \underline{\xi}(x)u + \underline{e}^T \underline{w} + \frac{1}{\gamma_1} tr(\dot{\Phi}_f \Phi_f^T)$$

$$+ \frac{1}{\gamma_2} tr(\dot{\Phi}_g \Phi_g^T)$$

$$= -\alpha|\underline{e}|^2 + \underline{e}^T \underline{w} + \frac{1}{\gamma_1} tr[(\gamma_1 \underline{e} \underline{\xi}^T(x) - \dot{\Theta}_f) \Phi_f^T]$$

$$+ \frac{1}{\gamma_2} tr\left[(\gamma_2 \underline{e} \underline{\xi}^T(x)u - \dot{\Theta}_g) \Phi_g^T\right]$$

$$= -\alpha|\underline{e}|^2 + \underline{e}^T \underline{w} + I_1^* \frac{\underline{e}^T \Theta_f \underline{\xi}(x)}{tr(\Theta_f \Theta_f^T)} tr(\Theta_f \Phi_f^T)$$

$$+ I_2^* \frac{\underline{e}^T \Theta_g \underline{\xi}(x)u}{tr(\Theta_g \Theta_g^T)} tr(\Theta_g \Phi_g^T) \quad (A.30)$$

where $I_1^* = 1$ ($I_2^* = 1$) if the second line of (11.26) ((11.27)) is true, $I_1^* = 0$ ($I_2^* = 0$) if the first line of (11.26) ((11.27)) is true, and we use the fact that

$\dot{\Phi}_f = -\dot{\Theta}_f$. We now prove that $I_1^* \frac{\underline{e}^T \Theta_f \underline{\xi}(x)}{tr(\Theta_f \Theta_f^T)} tr(\Theta_f \Phi_f^T) \leq 0$. If $I_1^* = 0$, the conclusion is trivial. For $I_1^* = 1$, which means that $tr(\Theta_f \Theta_f^T) = M_f$ and $\underline{e}^T \Theta_f \underline{\xi}(x) > 0$, we have

$$tr(\Theta_f \Phi_f^T) = tr\left[\Theta_f^* \Phi_f^T - \frac{1}{2}\Phi_f \Phi_f^T - \frac{1}{2}\Phi_f \Phi_f^T\right]$$

$$= \frac{1}{2} tr(\Theta_f^* \Theta_f^{*T}) - \frac{1}{2} tr(\Theta_f \Theta_f^T) - \frac{1}{2} tr(\Phi_f \Phi_f^T) \leq 0 \quad (A.31)$$

because $tr(\Theta_f^* \Theta_f^{*T}) \leq M_f = tr(\Theta_f \Theta_f^T)$ and $tr(\Phi_f \Phi_f^T) \geq 0$. Hence, $I_1^* \frac{\underline{e}^T \Theta_f \underline{\xi}(x)}{tr(\Theta_f \Theta_f^T)} tr(\Theta_f \Phi_f^T) \leq 0$. Similarly, we can prove that $I_2^* \frac{\underline{e}^T \Theta_g \underline{\xi}(x) u}{tr(\Theta_g \Theta_g^T)} tr(\Theta_g \Phi_g^T) \leq 0$. Therefore, from (A.30) we have

$$\dot{V} \leq -\alpha |\underline{e}|^2 + \underline{e}^T \underline{w} \leq -\alpha |\underline{e}|^2 + w_0 |\underline{e}| \quad (A.32)$$

(recall that $w_0 \equiv \sup_{t \geq 0} |\underline{w}(t)|$ which is assumed to be finite). Hence, if $|\underline{e}| > \frac{w_0}{\alpha}$, we have $\dot{V} < 0$, which means that $\underline{e} \in L_\infty$. Since $\underline{e} = \underline{x} - \hat{\underline{x}}$ and $\underline{x}$ is bounded by assumption, we have $\hat{\underline{x}} \in L_\infty$.

2. From (A.32) we have

$$\dot{V} \leq -\frac{\alpha}{2}|\underline{e}|^2 - \frac{\alpha}{2}\left[|\underline{e}|^2 - \frac{2}{\alpha}\underline{e}^T \underline{w} + \frac{1}{\alpha^2}|\underline{w}|^2 - \frac{1}{\alpha^2}|\underline{w}|^2\right]$$

$$\leq -\frac{\alpha}{2}|\underline{e}|^2 + \frac{1}{2\alpha}|\underline{w}|^2 \quad (A.33)$$

Integrating both sides of (A.33), we have

$$\int_0^t |\underline{e}(\tau)|^2 d\tau \leq \frac{2}{\alpha}(|V(0)| + |V(t)|) + \frac{1}{\alpha^2} \int_0^t |\underline{w}(\tau)|^2 d\tau \quad (A.34)$$

Define $a \equiv \frac{1}{\alpha}(|V(0)| + \sup_{t \geq 0} |V(t)|)$ and $b \equiv 1/\alpha^2$, (A.34) becomes (11.30). a is finite because $\underline{e}, tr(\Phi_f \Phi_f^T)$ and $tr(\Phi_g \Phi_g^T)$ are bounded from Part 1 of this theorem; therefore, $V \in L_\infty$ (see (A.28)).

3. If $\underline{w} \in L_2$, then from (11.30) we have that $\underline{e} \in L_2$. Because $\underline{e}, \Phi_f, \Phi_g, \underline{\xi}(x), u$, and $\underline{w}$ are all bounded ($\underline{\xi}(x)$ is bounded because from (11.22) we see that each element of $\underline{\xi}(x)$ is not greater than 1), from (A.29) we have that $\dot{\underline{e}} \in L_\infty$. Using the Barbalat's Lemma [Sastry and Bodson, 63] (if $\underline{e} \in L_2 \cap L_\infty$ and $\dot{\underline{e}} \in L_\infty$, then $\lim_{t \to \infty} |\underline{e}(t)| = 0$), we have $\lim_{t \to \infty} |\underline{e}(t)| = 0$. Q.E.D.

Proof of Theorem 11.3.

1. From (11.45) and (11.46) we see that if $|\underline{\theta}_f|^2 = M_f$, then: if $\underline{e}^T (\frac{\partial \hat{f}}{\partial \underline{\theta}_f})^T \underline{\theta}_f \leq 0$, we have that $\frac{d}{dt}(|\underline{\theta}_f|^2) = 2\underline{\dot{\theta}}_f^T \underline{\theta}_f = 2\gamma_1 \underline{e}^T (\frac{\partial \hat{f}}{\partial \underline{\theta}_f})^T \underline{\theta}_f \leq 0$; and if $\underline{e}^T (\frac{\partial \hat{f}}{\partial \underline{\theta}_f})^T \underline{\theta}_f$

> 0, we have that $\frac{d}{dt}(|\underline{\theta}_f|^2) = 2(\gamma_1 \underline{e}^T (\frac{\partial \hat{f}}{\partial \underline{\theta}_f})^T \underline{\theta}_f - \gamma_1 \frac{\underline{e}^T (\frac{\partial \hat{f}}{\partial \underline{\theta}_f})^T \underline{\theta}_f}{|\underline{\theta}_f|^2} \underline{\theta}_f^T \underline{\theta}_f) = 0$.
Hence, $|\underline{\theta}_f|^2 \leq M_f$. Using the same procedure we can prove that $|\underline{\theta}_g|^2 \leq M_g$. To prove $\hat{\underline{x}} \in L_\infty$, define the Lyapunov function candidate

$$V_1 = \frac{1}{2}|\underline{e}|^2 + \frac{1}{2\gamma_1}|\underline{\phi}_f|^2 + \frac{1}{2\gamma_2}|\underline{\phi}_g|^2 \qquad (A.35)$$

where $\underline{\phi}_f \equiv \underline{\theta}_f^* - \underline{\theta}_f$ and $\underline{\phi}_g \equiv \underline{\theta}_g^* - \underline{\theta}_g$. Using (11.12) and (11.16) (replace Θ_f^* and Θ_g^* by $\underline{\theta}_f^*$ and $\underline{\theta}_g^*$), we obtain the error dynamic equation

$$\dot{\underline{e}} = -\alpha \underline{e} + \left[\hat{f}(\underline{x}|\underline{\theta}_f^*) - \hat{f}(\underline{x}|\underline{\theta}_f)\right] + \left[\hat{g}(\underline{x}|\underline{\theta}_g^*) - \hat{g}(\underline{x}|\underline{\theta}_g)\right]u + w \qquad (A.36)$$

Using the Taylor series expansions of $\hat{f}(\underline{x}|\underline{\theta}_f^*)$ and $\hat{g}(\underline{x}|\underline{\theta}_g^*)$ around $\underline{\theta}_f$ and $\underline{\theta}_g$, we have

$$\hat{f}(\underline{x}|\underline{\theta}_f^*) - \hat{f}(\underline{x}|\underline{\theta}_f) = \left(\frac{\partial \hat{f}(\underline{x}|\underline{\theta}_f)}{\partial \underline{\theta}_f}\right)^T \underline{\phi}_f + \hat{f}_0 \qquad (A.37)$$

$$\hat{g}(\underline{x}|\underline{\theta}_g^*) - \hat{g}(\underline{x}|\underline{\theta}_g) = \left(\frac{\partial \hat{g}(\underline{x}|\underline{\theta}_g)}{\partial \underline{\theta}_g}\right)^T \underline{\phi}_g + \hat{g}_0 \qquad (A.38)$$

where $\hat{f}_0$ and $\hat{g}_0$ are the higher-order terms. Using the Mean Value Theorem, we have

$$\hat{f}_0 \equiv \hat{f}(\underline{x}|\underline{\theta}_f^*) - \hat{f}(\underline{x}|\underline{\theta}_f) - \left(\frac{\partial \hat{f}(\underline{x}|\underline{\theta}_f)}{\partial \underline{\theta}_f}\right)^T \underline{\phi}_f$$

$$= \left[\frac{\partial \hat{f}(\underline{x}|\underline{\theta}_{f0})}{\partial \underline{\theta}_f} - \frac{\partial \hat{f}(\underline{x}|\underline{\theta}_f)}{\partial \underline{\theta}_f}\right]^T \underline{\phi}_f \qquad (A.39)$$

$$\hat{g}_0 \equiv \hat{g}(\underline{x}|\underline{\theta}_g^*) - \hat{g}(\underline{x}|\underline{\theta}_g) - \left(\frac{\partial \hat{g}(\underline{x}|\underline{\theta}_g)}{\partial \underline{\theta}_g}\right)^T \underline{\phi}_g$$

$$= \left[\frac{\partial \hat{g}(\underline{x}|\underline{\theta}_{g0})}{\partial \underline{\theta}_g} - \frac{\partial \hat{g}(\underline{x}|\underline{\theta}_g)}{\partial \underline{\theta}_g}\right]^T \underline{\phi}_g \qquad (A.40)$$

where $\underline{\theta}_{f0} = \lambda_1 \underline{\theta}_f + (1 - \lambda_1)\underline{\theta}_f^*$ and $\underline{\theta}_{g0} = \lambda_2 \underline{\theta}_g + (1 - \lambda_2)\underline{\theta}_g^*$ for some $\lambda_1, \lambda_2 \in [0, 1]$. Substituting (A.37) and (A.38) into (A.36), we have

$$\dot{\underline{e}} = -\alpha \underline{e} + \left(\frac{\partial \hat{f}(\underline{x}|\underline{\theta}_f)}{\partial \underline{\theta}_f}\right)^T \underline{\phi}_f + \left(\frac{\partial \hat{g}(\underline{x}|\underline{\theta}_g)}{\partial \underline{\theta}_g}\right)^T \underline{\phi}_g u + \hat{f}_0 + \hat{g}_0 u + w \qquad (A.41)$$

The time derivative of V_1 along the solution of (A.41) is

$$\dot{V}_1 = -\alpha|\underline{e}|^2 + \underline{e}^T\left(\frac{\partial \hat{f}}{\partial \underline{\theta}_f}\right)^T \underline{\phi}_f + \underline{e}^T\left(\frac{\partial \hat{g}}{\partial \underline{\theta}_g}\right)^T \underline{\phi}_g u + \underline{e}^T \hat{f}_0 + \underline{e}^T \hat{g}_0 u + \underline{e}^T \underline{w}$$

$$+ \frac{1}{\gamma_1}\underline{\dot{\phi}}_f^T \underline{\phi}_f + \frac{1}{\gamma_2}\underline{\dot{\phi}}_g^T \underline{\phi}_g \tag{A.42}$$

Using (11.45)–(11.48) and $\underline{\dot{\phi}}_f = -\underline{\dot{\theta}}_f$, $\underline{\dot{\phi}}_g = -\underline{\dot{\theta}}_g$, we have that

$$\dot{V}_1 = -\alpha|\underline{e}|^2 + \underline{e}^T \hat{f}_0 + \underline{e}^T \hat{g}_0 u + \underline{e}^T \underline{w} + I_1^* \frac{\underline{e}^T(\frac{\partial \hat{f}}{\partial \underline{\theta}_f})^T \underline{\theta}_f}{|\underline{\theta}_f|^2} \underline{\theta}_f^T \underline{\phi}_f$$

$$+ I_2^* \frac{\underline{e}^T(\frac{\partial \hat{g}}{\partial \underline{\theta}_g})^T \underline{\theta}_g u}{|\underline{\theta}_g|^2} \underline{\theta}_g^T \underline{\phi}_g \tag{A.43}$$

where $I_1^* = 0$ (1) if the condition of the first (second) line of (11.45) is true, and $I_2^* = 0$ (1) if the condition of the first (second) line of (11.46) is true. Using the same arguments as in the proof of Theorem 11.2 and noticing that $|\underline{\theta}|^2 = tr(\underline{\theta}\underline{\theta}^T)$, we can prove that the last two terms of (A.43) are nonpositive. Hence,

$$\dot{V}_1 \leq -\alpha|\underline{e}|^2 + \underline{e}^T \hat{f}_0 + \underline{e}^T \hat{g}_0 u + \underline{e}^T \underline{w}$$

$$\leq -\alpha|\underline{e}|^2 + |\underline{e}| \cdot |\hat{f}_0| + |\underline{e}| \cdot |\hat{g}_0| \cdot |u| + |\underline{e}| \cdot |\underline{w}| \tag{A.44}$$

From (11.39)–(11.44) we see that all the elements of $\frac{\partial \hat{f}}{\partial \underline{\theta}}$ and $\frac{\partial \hat{g}}{\partial \underline{\theta}}$ are bounded; therefore, from (A.39) and (A.40) and the boundness of $\underline{\phi}_f$ and $\underline{\phi}_g$ (we have proven that $|\underline{\theta}_f|^2 \leq M_f$ and $|\underline{\theta}_g|^2 \leq M_g$; $|\underline{\theta}_f^*|^2 \leq M_f$ and $|\underline{\theta}_g^*|^2 \leq M_g$ by definition), we have that $|\hat{f}_0|$ and $|\hat{g}_0|$ are bounded. Hence, $\beta \equiv |\hat{f}_0| + |\hat{g}_0| \cdot |u| + |\underline{w}|$ is finite. From (A.44) we have that if $|\underline{e}| > \beta/\alpha$, then $\dot{V}_1 < 0$; therefore, $|\underline{e}|$ is bounded. Since $\hat{\underline{x}} = \underline{x} - \underline{e}$ and $\underline{x}$ is bounded by assumption, we have $\hat{\underline{x}} \in L_\infty$.

2. From (A.39), (A.40), (11.50), and the first line of (A.44), we have that

$$\dot{V}_1 \leq -\frac{\alpha}{2}|\underline{e}|^2 - \frac{\alpha}{2}\left(|\underline{e}|^2 - \frac{2}{\alpha}\underline{e}^T \underline{v} + \frac{1}{\alpha^2}|\underline{v}|^2 - \frac{1}{\alpha^2}|\underline{v}|^2\right)$$

$$\leq -\frac{\alpha}{2}|\underline{e}|^2 + \frac{1}{2\alpha}|\underline{v}|^2 \tag{A.45}$$

Integrating both sides of (A.45) and defining $a \equiv \frac{1}{\alpha}(|V_1(0)| + sup_{t\geq 0}|V(t)|)$ and $b \equiv 1/\alpha^2$, we obtain (11.49) ($sup_{t\geq 0}|V(t)|$ is finite because $\underline{e}, \underline{\phi}_f,$ and $\underline{\phi}_g$ are all bounded).

3. If $\underline{v} \in L_2$, then from (11.49) $\underline{e} \in L_2$. From Part 1 of this theorem, $\underline{e} \in L_\infty$. Because in the proof of Part 2 of this theorem we showed that all the variables in the right-hand side of (A.41) are bounded, we have $\underline{\dot{e}} \in L_\infty$. Using the Barbalat's Lemma, we have $lim_{t \to \infty} |\underline{e}(t)| = 0$. Q.E.D.

REFERENCES

[1] Bellman, R. E. and L. A. Zadeh, "Local and fuzzy logics," in *Modern Uses of Multiple-Valued Logic*, J. M. Dunn and G. Epstein, eds. Reidel Publ., Dordrecht, Netherlands: Reidel Publ., 1977, pp. 103–165.

[2] Bernard, J. A., "Use of rule-based system for process control," *IEEE Contr. Syst. Mag.*, 8, no. 5 (1988), 3–13.

[3] Biglieri, E., A. Gersho, R. D. Gitlin, and T. L. Lim, "Adaptive cancellation of nonlinear intersymbol interference for voiceband data transmission," *IEEE J. on Selected Areas in Communications*, SAC-2, no. 5 (1984), 765–777.

[4] Box, G.E.P. and G. M. Jenkins, *Time Series Analysis: Forecasting and Control*, London: Holden-Day Inc., 1976.

[5] Chen, S., C.F.N. Cowan, and P. M. Grant, "Orthogonal least squares learning algorithm for radial basis function networks," *IEEE Trans. on Neural Networks*, 2, no. 2 (1991), 302–309.

[6] Chen, S., S. A. Billings, and W. Luo, "Orthogonal least squares methods and their application to non-linear system identification," *Int. J. Contr.*, 50, no. 5 (1989), 1873–1896.

[7] Chen, S., G. J. Gibson, C.F.N. Cowan, and P. M. Grant, "Adaptive equalization of finite non-linear channels using multilayer perceptrons," *Signal Processing*, 20 (1990), 107–119.

[8] Chen, S., G. J. Gibson, C.F.N. Cowan, and P. M. Grant, "Reconstruction of binary signals using an adaptive radial-basis-function equalizer," *Signal Processing*, 22 (1991), 77–93.

[9] Chiu, S., S. Chand, D. Moore, and A. Chaudhary, "Fuzzy logic for control of roll and moment for a flexible wing aircraft," *IEEE Control Systems Magazine*, 11, no. 4 (1991), 42–48.

[10] Ciliz, K., J. Fei, K. Usluel, and C. Isik, "Practical aspects of the knowledge-based control of a mobile robot motion," *Proc. 30th Midwest Symp. on Circuits and Systems*, Syracuse, NY, 1987.

[11] Cowan, C.F.N. and P. M. Grant (eds.), *Adaptive Filters*, Englewood Cliffs, NJ: Prentice-Hall, Inc., 1985.

[12] Cybenko, G. "Approximation by superpositions of a sigmoidal function," *Mathematics of Control, Signals, and Systems*, no. 2, (1989), 303–314.

[13] Dubois, D. and H. Prade, *Fuzzy Sets and Systems: Theory and Applications*. Orlando, FL: Academic Press, Inc., 1980.

[14] Falconer, D. D., "Adaptive equalization of channel nonlinearities in QAM data transmission systems," *The Bell System Technical Journal*, 57, no. 7 (1978), 2589–2611.

[15] Fujitec, F., "FLEX-8000 series elevator group control system," Fujitec Co., Ltd., Osaka, Japan, 1988.

[16] Goodwin, G. C. and R. L. Payne, *Dynamic System Identification: Experiment Design and Data Analysis*. New York: Academic Press, Inc., 1977.

[17] Hale, J. K., *Ordinary Differential Equations*. New York: Wiley Inter Science, 1969.

[18] Hauser, J., S. Sastry, and P. Kokotovic, "Nonlinear control via approximate input-output linearization: the ball and beam example," *IEEE Trans. on Automatic Control*, AC-37, no. 3 (1992), 392–398.

[19] Holden, A. V., *Chaos*. Princeton, NJ: Princeton University Press, 1986.

[20] Hornik, K., M. Stinchcombe, and H. White, "Multilayer feedforward networks are universal approximators," *Neural Networks*, no. 2 (1989), 359–366.

[21] Isidori, A., *Nonlinear Control Systems*. Berlin: Springer-Verlag, 1989.

[22] Itoh, O., K. Gotoh, T. Nakayama, and S. Takamizawa, "Application of fuzzy control to activated sludge process," *Proc. 2nd IFSA Congress*. Tokyo, Japan (1987), 282–285.

[23] Kasai, Y. and Y. Morimoto, "Electronically controlled continuously variable transmission," *Proc. Int. Congress on Transportation Electronics*. Dearborn, MI, 1988.

[24] Kickert, W. J. M. and H. R. Van Nauta Lemke, "Application of a fuzzy controller in a warm water plant," *Automatica*, 12, no. 4 (1976), 301–308.

[25] Kinoshita, M., T. Fukuzaki, T. Satoh, and M. Miyake, "An automatic operation method for control rods in BWR plants," *Proc. Specialists' Meeting on In-Core Instrumentation and Reactor Core Assessment*, Cadarache, France, 1988.

[26] Kiszka, J., M. Gupta, and P. Nikiforuk, "Energetistic stability of fuzzy dynamic systems," *IEEE Trans. Systems, Man, and Cybern.*, SMC-15, no. 5 (1985), 783–792.

[27] Klir, G. J. and T. A. Folger, *Fuzzy Sets, Uncertainty, and Information*, Englewood Cliffs, NJ: Prentice Hall, 1988.

[28] Kong, S. G. and B. Kosko, "Comparison of Fuzzy and Neural Truck Backer-Upper Control Systems," *Proc. IJCNN-90*, 3 (June 1990), 349–358.

[29] Langari, G. and M. Tomizuka, "Stability of fuzzy linguistic control systems," *Proc. IEEE Conf. on Decision and Control* (1990), 2185–2190.

[30] Lapedes, A. and R. Farber, "Nonlinear signal processing using neural networks: prediction and system modeling," *LA-UR-87-2662*, 1987.

References

[31] Larkin, L. I., "A fuzzy logic controller for aircraft flight control," in *Industrial Applications of Fuzzy Control*, ed. M. Sugeno, Amsterdam: North-Holland, (1985), 87–104.

[32] Larsen, P. M., "Industrial application of fuzzy logic control," *Int. J. Man Mach. Studies*, 12, no. 1 (1980), 3–10.

[33] Lee, C. C., "Fuzzy logic in control systems: fuzzy logic controller, part I," *IEEE Trans. on Syst., Man, and Cybern.*, SMC-20, no. 2 (1990), 404–418.

[34] Lee, C. C., "Fuzzy logic in control systems: fuzzy logic controller, part II," *IEEE Trans. on Syst., Man, and Cybern.*, SMC-20, no. 2 (1990), 419–435.

[35] Lindley, D. V., "The probability approach to the treatment of uncertainty in artificial intelligence and expert systems," *Statistical Science*, 2, no. 1 (1987), 17–24.

[36] Lippmann, R., "A critical overview of neural network pattern classifiers," *Proc. 1991 IEEE Workshop on Neural Networks for Signal Processing*, Princeton, NJ (1991), 266–275.

[37] Ljung, L., *System Identification—Theory for the User*. Englewood Cliffs, NJ: Prentice-Hall, 1987.

[38] Ljung, L., "Issues in system identification," *IEEE Control Systems Magazine* (January 1991), 25–29.

[39] Luenberger, D. G., *Linear and Nonlinear Programming*. Reading, MA: Addison-Wesley Publishing Company, Inc., 1984.

[40] Maiers, J. and Y. S. Sherif, "Applications of fuzzy sets theory," *IEEE Trans. Syst. Man Cybern.*, SMC-15, no. 6 (1985), 175–189.

[41] Mamdani, E. H., "Applications of fuzzy algorithms for simple dynamic plant," *Proc. IEE*, 121, no. 12 (1974), 1585–1588.

[42] Mamdani, E. H. and S. Assilian, "An experiment in linguistic synthesis with a fuzzy logic controller," *Int. J. Man Mach. Studies*, 7, no. 1 (1975), 1–13.

[43] Mathews, V. J., "Adaptive polynomial filters," *IEEE Signal Processing Magazine* (July 1991), 10–26.

[44] Meisel, W. S., "Potential functions in mathematical pattern recognition," *IEEE Trans. on Computers*, C-18, no. 10 (1969), 911–918.

[45] Miller, W. T., R. S. Sutton, and P. J. Werbos, eds., *Neural Networks for Control*. Cambridge, MA: MIT Press, 1990.

[46] Minsky, M. and S. Papert, *Perceptron: An Introduction to Computational Geometry*, Cambridge, MA: MIT Press, 1969.

[47] Moody, J. and C. Darken, "Learning with localized receptive fields," in *Proc. 1988 Connectionist Models Summer School*, Carnegie Mellon University, 1988.

[48] Narendra, K. S. and A. M. Annaswamy, *Stable Adaptive Systems*, Englewood Cliffs, NJ: Prentice-Hall, 1989.

[49] Narendra, K. S. and K. Parthasarathy, "Identification and control of dynamical systems using neural networks," *IEEE Trans. on Neural Networks*, 1, no. 1 (1990), 4–27.

[50] Naylor, A. W. and G. R. Sell, *Linear Operator Theory in Engineering and Science*. New York: Springer-Verlag, 1982.

[51] Nguyen, D. and B. Widrow, "The truck backer-upper: an example of self-learning in neural networks," *IEEE Cont. Syst. Mag.*, 10, no. 3 (1990), 18–23.

[52] Ostergaad, J. J., "Fuzzy logic control of a heat exchange process," in *Fuzzy Automata and Decision Processes*, eds. M. M. Gupta, G. N. Saridis, and B. R. Gaines. Amsterdam: North-Holland, (1977), 285–320.

[53] Pappis, C. P. and E. H. Mamdani, "A fuzzy logic controller for a traffic junction," *IEEE Trans. Syst. Man Cybern.*, SMC-7, no.10 (1977), 707–717.

[54] Parzen, E., "On estimation of a probability density function and mode," *Ann. Math. Statist.*, 33 (1962), 1065–1076.

[55] Pedrycz, W., *Fuzzy Control and Fuzzy Systems*. New York: John Wiley & Sons, Inc., 1989.

[56] Pitas, I. and A. N. Venetsanopoulos, *Nonlinear Digital Filters*. Boston: Kluwer Academic Publishers, 1990.

[57] Polycarpou, M. M. and P. A. Ioannou, "Identification and control of nonlinear systems using neural network models: design and stability analysis," USC EE-Report, 1991.

[58] Powell, M. J. D., *Approximation Theory and Methods*. Cambridge, MA: Cambridge University Press, 1981.

[59] Powell, M. J. D., "Radial basis functions for multivariable interpolation: A review," in *Algorithms for Approximation*, eds., J. C. Mason and M. G. Cox. Oxford: Oxford University Press, 1987, 143–167.

[60] Rudin, W., *Principles of Mathematical Analysis*. New York. McGraw-Hill, Inc., 1976.

[61] Rumelhart, D. E. and J. L. McCleland, eds., *Parallel Distributed Processing I, II*. Cambridge, MA: MIT Press, 1986.

[62] Sanner, R. M. and J. E. Slotine, "Gaussian networks for direct adaptive control," *Proc. American Control Conf.* (1991), 2153–2159.

[63] Sastry, S. and M. Bodson, *Adaptive Control: Stability, Convergence, and Robustness*. Englewood Cliffs, NJ: Prentice-Hall, 1989.

[64] Sastry, S. and A. Isidori, "Adaptive control of linearizable systems," *IEEE Trans. on Automatic Control*, 34, no. 11, 1989, 1123–1131.

[65] Schwartz, A. L., "Comments on 'Fuzzy logic for control of roll and moment for a flexible wing aircraft,'" *IEEE Control Systems Magazine*, 12, no. 1 (Febuary 1992), 61–62.

[66] Slotine, J. E. and W. Li, *Applied Nonlinear Control*. Englewood Cliffs, NJ: Prentice-Hall, Inc., 1991.

[67] Specht, D. F., "A general regression neural network," *IEEE Trans. on Neural Networks*, 2, no. 6 (1991), 568–576.

[68] Specht, D. F., "Generation of polynomial discriminant functions for pattern recognition," *IEEE Trans. Electron. Comput.*, EC-16 (1967), 308–319.

[69] Stallings, W., "Fuzzy set theory versus Bayesian statistics," *IEEE Trans. Systems, Man, and Cybern.*, 7, no. 3 (1977), 216–219.

[70] Sugeno, M. and M. Nishida, "Fuzzy control of model car," *Fuzzy Sets Syst.*, 16 (1985), 103–113.

[71] Takagi, H., "Introduction to Japanese consumer products that apply neural networks and fuzzy systems," Personal communication, 1992.

[72] Takagi, T. and M. Sugeno, "Fuzzy identification of systems and its applications to modeling and control," *IEEE Trans. on Systems, Man, and Cybern.*, SMC-15(1) (1985), 116–132.

[73] Tong, R. M., "Some properties of fuzzy feedback systems," *IEEE Trans. Systems, Man, and Cybern.*, SMC-10, no. 6 (1980), 327–330.

[74] Tong, R. M., M. B. Beck, and A. Latten, "Fuzzy control of the activated sludge wastewater treatment process," *Automatica*, 16, no.6 (1980), 695–701.

[75] Togai, M. and H. Watanabe, "Expert system on a chip: an engine for real-time approximate reasoning," *IEEE Expert Syst. Mag.*, 1 (1986), 55–62.

[76] Togai, M. and S. Chiu, "A fuzzy accelerator for a programming environment for real-time fuzzy control," *Proc. 2nd IFSA Congress*, Tokyo, Japan (1987), 147–151.

[77] Umbers, I. G. and P. J. King, "An analysis of human decision making in cement kiln control and the implications for automation," *Int. J. Man Mach. Studies*, 12, no. 1 (1980), 11–23.

[78] Uragami, M., M. Mizumoto and K. Tananka, "Fuzzy robot controls," *Cybern.* 6 (1976), 39–64.

[79] Wang, L. X., "Fuzzy systems are universal approximators," *Proc. IEEE International Conf. on Fuzzy Systems*, San Diego (1992), 1163–1170.

[80] Wang, L. X., "Stable adaptive fuzzy control of nonlinear systems," *Proc. 31st IEEE Conf. on Decision and Control*, Tucson (1992), 2511–2516.

[81] Wang, L. X., "Fuzzy systems as nonlinear dynamic system identifiers: Part I and II," *Prof. 31st IEEE Conf. on Decision and Control*, Tucson (1992), 897–902 and 3418–3422.

[82] Wang, L. X. and J. M. Mendel, "Generating fuzzy rules by learning from examples," *Proc. 6th IEEE International Symposium on Intelligent Control* (1991), 263–268.

[83] Wang, L. X. and J. M. Mendel, "Generating fuzzy rules from numerical data, with applications," USC SIPI Report, no. 169, 1991; also, *IEEE Trans. on Systems, Man, and Cybern.*, 22, no. 6 (1992), 1414–1427.

[84] Wang, L. X. and J. M. Mendel, "Analysis and design of fuzzy logic controller," USC SIPI Report, no. 184, 1991.

[85] Wang, L. X. and J. M. Mendel, "Fuzzy basis functions, universal approximation, and orthogonal least squares learning," *IEEE Trans. on Neural Networks*, 3, no. 5 (1992), 807–814.

[86] Wang, L. X. and J. M. Mendel, "Back-propagation fuzzy systems as nonlinear dynamic system identifiers," *Proc. IEEE International Conf. on Fuzzy Systems*, San Diego (1992), 1409–1418.

[87] Wang. L. X. and J. M. Mendel, "Structured trainable networks for matrix algebra," *Proc. 1990 International Joint Conf. on Neural Networks*, 2 (1990), II125–II132.

[88] Wang, L. X. and J. M. Mendel, "Matrix computations and equation solving using structured networks and training," *Proc. IEEE 1990 Conf. on Decision and Control* (1990), 1747–1750.

[89] Wang, L. X. and J. M. Mendel, "Parallel structured network for solving a wide variety of matrix algebra problems," *Journal of Parallel and Distributed Computing*, 14 (1992), 236–247.

[90] Wang, L. X. and J. M. Mendel, "Three-dimensional structured networks for matrix equation solving," *IEEE Trans. on Computers*, 40, no. 12 (1991), 1337–1346.

[91] Wang, L. X. and J. M. Mendel, "Cumulant-based parameter estimation using structured networks," *IEEE Trans. on Neural Networks*, 2, no. 1 (1991), 73–83.

[92] Wang, L. X. and J. M. Mendel, "Adaptive prediction-error deconvolution and source wavelet estimation using Hopfield neural networks," *Geophysics*, 57, no. 5 (1992), 670–679.

[93] Wang, L. X. and J. M. Mendel, "Fuzzy adaptive filters, with application to nonlinear channel equalization," *IEEE Trans. on Fuzzy Systems*, 1, no. 3, 1993.

[94] Watanabe, H. and W. Dettloff, "Reconfigurable fuzzy logic processor: a full custom digital VLSI," *Proc. Int. Workshop on Fuzzy System Applications*, Iizuka, Japan (1988), 49–50.

[95] Werbos, P., "New Tools for Predictions and Analysis in the Behavioral Science," Ph.D. Thesis, Harvard University Committee on Applied Mathematics, 1974.

[96] Widrow, B. and S. D. Stearns, *Adaptive Signal Processing*. Englewood Cliffs, NJ: Prentice-Hall, 1984.

[97] Yamakawa, T. and T. Miki, "The current mode fuzzy logic integrated circuits fabricated by standard CMOS process," *IEEE Trans. Computer*, C-35, no. 2 (1986), 161–167.

[98] Yamakawa, T. and K. Sasaki, "Fuzzy memory device," *Proc. 2nd IFSA Congress*, Tokyo, Japan (1987), 551–555.

[99] Yamakawa, T., "Fuzzy microprocessors—rule chip and defuzzification chip," *Proc. Int. Workshop on Fuzzy System Applications* (1988), 51–52.

[100] Yagishita, O., O. Itoh, and M. Sugeno, "Application of fuzzy reasoning to the water purification process," in *Industrial Applications of Fuzzy Control*, ed. M. Sugeno. Amsterdam: North-Holland, (1985), 19–40.

[101] Yasunobu, S., S. Miyamoto, and H. Ihara, "Fuzzy control for automatic train operation system," *Proc. 4th IFAC/IFIP/IFORS Int. Congress on Control in Transportation Systems*, Baden-Baden, 1983.

[102] Yasunobu, S. and S. Miyamoto, "Automatic train operation by predictive fuzzy control," in *Industrial Application of Fuzzy Control*, ed. M. Sugeno, Amsterdam: North-Holland, (1985), 1–18.

[103] Yasunobu, S., S. Sekino, and T. Hasegawa, "Automatic train operation and automatic crane operation systems based on predictive fuzzy control," *Proc. 2nd IFSA Congress*, Tokyo, Japan (1987), 835–838.

[104] Yasunobu, S. and T. Hasegawa, "Evaluation of an automatic container crane operation system based on predictive fuzzy control," *Control Theory Adv. Technol.*, 2, no. 3 (1986), 419–432.

[105] Yasunobu, S. and T. Hasegawa, "Predictive fuzzy control and its application for automatic container crane operation system," *Proc. 2nd IFSA Congress*, Tokyo, Japan (1987), 349–352.

[106] Zadeh, L. A., "Fuzzy sets," *Informat. Control*, 8 (1965), 338–353.

[107] Zadeh, L. A., "Toward a theory of fuzzy systems," in *Aspects of Network and System Theory*, eds. R. E. Kalman and N. DeClaris, New York: Rinehart and Winston, 1971.

[108] Zadeh, L. A., "Outline of a new approach to the analysis of complex systems and decision processes," *IEEE Trans. on Systems, Man, and Cybern.*, SMC-3, no. 1 (1973), 28–44.

References

[109] Zadeh, L. A., "A theory of approximating reasoning," in *Machine Intelligence*, 9, eds., J. E. Hayes, D. Michie, and L. I. Mikulich. New York: Elsevier (1979), 149–194.

[110] Zadeh, L. A., "Syllogistic reasoning in fuzzy logic and its application to usuality and reasoning with dispositions," *IEEE Trans. on Systems, Man, and Cybern.*, SMC-15, no. 6 (1985), 754–763.

INDEX

A

adaptive fuzzy control
    based on input-output linearization, 158, 159
    direct, 104, 140–48
    first-type, 106, 114–18, 144–46
    indirect, 104, 107–22
    second-type, 106, 118–22, 146–48
    why, 102, 103

adaptive fuzzy systems, 2

adaptive law, 110–12, 115, 116, 120, 121, 145, 147, 159

algebraic product, 10

algebraic sum, 11

approximate reasoning, 13

arithmetic rule of fuzzy implication, 18

artificial neural networks, 29

assumptions, 109, 113, 141, 144, 158, 165, 167

autocorrelation of residuals, 62

B

back-propagation
    for fuzzy logic systems, 29–31
    for multilayer perceptron, 95, 96

ball-and-beam system, 54–60, 160–62

Barbalat's Lemma, 214, 217, 220

Bayesian statisticians, 100

Bellman-Gronwall Lemma, 215

bit error rate, 200

Boolean rule of fuzzy implication, 18

bounded product, 10

bounded sum, 11

C

center of fuzzy set, 9, 68

certainty equivalent controller, 108, 109

chain rule, 30, 31

chaotic Duffing forced oscillation, 151

chaotic time series, 60

chi-squares test, 62, 63
cluster, 85, 86
complement, 10
conditional mean, 98
conditional probability function, 190
control objectives, 107

D

debate on fuzziness versus probability, 101
defuzzifier
    center average, 22, 24, 25
    maximum, 22
    modified center average, 22, 26
degree of
    membership, 9
    rules, 67
domain interval, 66
drastic product, 10
drastic sum, 11

E

error equations, 108, 141–43, 154
equalization of nonlinear channel, 189
expert control, 205, 206
extension principle, 11

F

feedforward networks, 30
fuzzifier, 5, 22
    nonsingleton, 22
    singleton, 22, 24–26
fuzziness versus probability, 101
fuzzy
    conjunction, 14
    disjunction, 14
    implication, 14
fuzzy adaptive filters, 182
    LMS, 186–89
    RLS, 183–86
fuzzy basis functions (FBF), 50, 106, 115, 184
fuzzy entropy, 204

fuzzy identifiers, 32, 164–81
    first-type, 166–71
    second-type, 171–75
    as universal approximator, 165, 166
fuzzy inference engine, 3, 17
fuzzy logic controllers, 102, 202, 203
fuzzy logic systems
    definition, 2
    as universal approximators, 27
    useful classes, 24–26
fuzzy relations, 11
fuzzy rule base, 3, 15
fuzzy set, 9
fuzzy singleton, 9

G

Gaussian function
    functions, 97
    membership functions, 25, 119
    noise, 191
generalization and matching, 84
generalized modus ponens (GMP), 13
generalized modus tollens (GMT), 13
geometric formulation of equalization, 189–91
geometric interpretation of projection, 118
Goguen's rule of fuzzy implication, 18
gradient metrices, 172, 173
gradual rule, 16

H

hedge, 12
hierarchical systems, 206
Hurwitz
    matrix, 165, 215
    polynomial, 55
hybrid systems, 205–209

I

identification of goals, 167
identification of nonlinear dynamic systems, 32–48, 164–81
incomplete-IF-part-rule, 15

Index

information
  linguistic, 1
  measure, 204
  numerical, 1
initial controller construction, 114, 115, 119, 144, 146
initial FBF determination, 52
initial parameters, 32, 35
initial weights and offsets, 95
input-output linearization, 155–57
internal dynamics, 157
intersection, 10
intelligent control, 205–209
intelligent vehicle/highway systems, 207, 209
inverted pendulum tracking control, 122–39

L

large-sample problems, 83, 85
least mean squares (LMS)
  fuzzy adaptive filter, 186–88
Lemmas, 24–26, 210–212
linear in the parameters, 50, 52, 106
linguistic information, 1, 36, 128, 194
linguistic variables, 12
Lipschitz condition, 214, 215
logistic functions, 97
Lyapunov
  equation, 108, 141
  function, 111, 142

M

material implication, 14
maximum rule
  of fuzzy implication, 18
Mean Value Theorem, 218
membership function, 9
modus ponens
  generalized, 13
  generalization of, 14
modus tollens
  generalized, 13
  generalization of, 14

N

nearest neighborhood clustering, 85, 86
neural identifiers, 32
non-fuzzy rules, 17
numerical data, 65
numerical information, 1

O

off-line processing, 114, 115, 144, 146
on-line adaptation, 115, 116, 120, 145, 147
optimal equalizer, 191
optimal fuzzy logic systems, 84
or rules, 16, 68
orthogonal decomposition, 54
orthogonal least squares, 52–54

P

parallel model, 33, 34
perceptron, 93–96
phase plane, 152, 153, 176
platoon, 208, 209
probabilistic general regression, 98–00
projection operators, 115, 118, 169, 173
propositional calculus, 14
pure fuzzy logic systems, 3

R

radial basis function (RBF), 96–98
radius, 85
recursive least squares (RLS)
  fuzzy adaptive filter, 183–85
regression, 52, 98
relative degree, 157, 160
residual sequence, 61
robot, 207, 208

S

semantic, 12
series-parallel model, 33, 167
signal-to-noise ratio, 200
small-sample problem, 83
stable (definition), 107
statistical tests, 60–63
Stone-Weierstrass Theorem, 210
supervisory controller, 109, 110, 141, 151, 153
support of fuzzy set, 9
sup-star composition, 3, 11, 17
symbolic logic, 85
syntactic, 12

T

table-lookup scheme, 65–69
Taylor series expansions, 112, 143, 218
T-conorm, 10
theorems, 27, 35, 84, 116, 121, 145, 147, 166, 170, 174
time-series prediction, 76–82
T-norm, 10
tracking error, 115
triangular membership functions, 26
truck backer-upper control, 69–76

U

union, 10
universal approximator, 27, 165, 166
unless rule, 16

W

weighted average, 4, 97
weighted sum, 97
why adaptive fuzzy control, 104
why fuzzy control, 102, 103